BIG
JOBS GUIDE

BIG JOBS GUIDE

Business, Industry, and Government Careers
for Mathematical Scientists, Statisticians,
and Operations Researchers

Rachel Levy
Harvey Mudd College
Claremont, California

Richard Laugesen
University of Illinois
Urbana, Illinois

Fadil Santosa
University of Minnesota
Minneapolis, Minnesota

Society for Industrial and Applied Mathematics
Philadelphia

Copyright © 2018 by the Society for Industrial and Applied Mathematics

10 9 8 7 6 5 4 3 2 1

All rights reserved. Printed in the United States of America. No part of this book may be reproduced, stored, or transmitted in any manner without the written permission of the publisher. For information, write to the Society for Industrial and Applied Mathematics, 3600 Market Street, 6th Floor, Philadelphia, PA 19104-2688 USA.

Publications Director	Kivmars H. Bowling
Acquisitions Editor	Paula Callaghan
Developmental Editor	Gina Rinelli Harris
Managing Editor	Kelly Thomas
Production Editor	Louis R. Primus
Copy Editor	T&T Productions Ltd
Compositor	T&T Productions Ltd
Production Manager	Donna Witzleben
Production Coordinator	Cally A. Shrader
Graphic Designer	Lois Sellers

Ampl is a registered trademark of AMPL Optimization LLC, Lucent Technologies Inc.
COMSOL is a registered trademark of COMSOL Inc.
Data Carpentry is a registered trademark of Data Carpentry.
Excel is a trademark of Microsoft Corporation in the United States and/or other countries.
GitHub is a registered trademark of GitHub, Inc
Java is a trademark of Sun Microsystems, Inc. in the United States and other countries.
Maple is a trademark of Waterloo Maple, Inc.
Mathematica is a registered trademark of Wolfram Research, Inc.
MATLAB is a registered trademark of The MathWorks, Inc. For MATLAB product information, please contact The MathWorks, Inc., 3 Apple Hill Drive, Natick, MA 01760-2098 USA, 508-647-7000, Fax: 508-647-7001, info@mathworks.com, www.mathworks.com.
OpenMP and the OpenMP logo are registered trademarks of the OpenMP Architecture Review Board in the United States and other countries. All rights reserved.
Software Carpentry and the Software Carpentry logo are registered trademarks of Community Initiatives.
Strong Interest Inventory is a trademark or registered trademark of CPP, Inc., in the United States and other countries.
Subversion is a registered trademark of The Apache Software Foundation.
Visual Basic is a registered trademark of Microsoft Corporation in the United States and/or other countries.

Library of Congress Cataloging-in-Publication Data
Names: Levy, Rachel, 1968- author. | Laugesen, Richard, 1967- author. | Santosa, Fadil, author.
Title: BIG jobs guide : business, industry, and government careers for mathematical scientists, statisticians, and operations researchers / Rachel Levy (Harvey Mudd College, Claremont, California), Richard Laugesen (University of Illinois, Urbana, Illinois), Fadil Santosa (University of Minnesota, Minneapolis, Minnesota).
Other titles: Business, industry, and government careers for mathematical scientists, statisticians, and operations researchers
Description: Philadelphia : Society for Industrial and Applied Mathematics, [2018] | Series: Other titles in applied mathematics ; 158 | Includes bibliographical references and index.
Identifiers: LCCN 2018015652 (print) | LCCN 2018018379 (ebook) | ISBN 9781611975291 | ISBN 9781611975284 (print)
Subjects: LCSH: Mathematicians--Employment. | Mathematicians--Vocational guidance. | Statisticians--Employment. | Statisticians--Vocational guidance.
Classification: LCC QA10.5 (ebook) | LCC QA10.5 .L48 2018 (print) | DDC 510.23--dc23
LC record available at https://lccn.loc.gov/2018015652

 is a registered trademark.

Contents

Abbreviations		vii
Foreword		ix
Preface		xi
1	Why the *BIG Jobs Guide* is for you	1
2	Why you need to think BIG	7
3	Who you are: your "special sauce"	17
4	What you should study	25
5	What to put in your résumé	45
6	Why do an internship?	57
7	What jobs are out there?	67
8	What it is like to take a BIG job	77
9	Ways to collect career mentors	87
10	Winning at the job search	95
11	What can departments do for BIG careers?	109
12	What international students must know	123
13	Ways to establish and mentor internships	127
14	Where to learn more	135
Index		139

Abbreviations

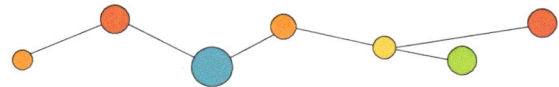

AMS — American Mathematical Society

ASA — American Statistical Association

BIG — Business, Industry, and Government

INFORMS — Institute for Operations Research and the Management Sciences

MAA — Mathematical Association of America

M3 Challenge — MathWorks Math Modeling Challenge

OR — Operations Research

PIC Math — Preparation for Industrial Careers in Mathematical Sciences

REU — Research Experiences for Undergraduates

SIAM — Society for Industrial and Applied Mathematics

Foreword

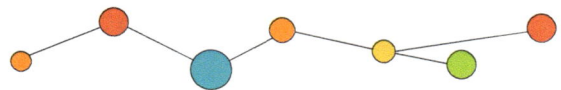

I am delighted by the appearance of this guide, advising budding mathematical scientists about how to approach the job search for positions in business, industry, and government (BIG). The authors give practical and compelling advice on how to navigate that territory, based on their deep and very successful experiences in advisory roles. They concentrate on students trained in mathematics, statistics, and operations research. Computer science students will also benefit from the fantastic career advice outlined by the authors. The main transitions considered are undergraduates envisioning BIG jobs, and graduate students as well as postdoctoral faculty contemplating BIG careers. The job-seeking principles provided will be helpful for other transitions too. The authors are to be congratulated for their eminently practical and outstanding work.

The mathematical sciences play an ever-increasing role throughout the sciences and the BIG world. Recent examples have occurred in energy technologies, genomics, finance, and data analysis. Novel techniques developed in the mathematical sciences have direct applications. Modern life, from search engines to aircraft design, from financial markets to medical imaging, has been enabled by mathematical science methodologies. That is why mathematical science skills will be in increasing demand for the foreseeable future—they lead to a fantastic variety of career opportunities, not yet captured by classical nomenclatures.

All fields of the mathematical sciences show up as useful. Higher degree levels generally lead to faster career progress and higher levels of responsibility. At the top level of responsibilities, a PhD is a de facto requirement. At all levels the "special sauce," in the terminology of Chapter 3, is made of mathematical thinking and problem-solving skills.

Best wishes for a great read, and a satisfying career deploying and developing your special talents.

Dr. Philippe Tondeur

Professor Emeritus, University of Illinois, Urbana–Champaign, U.S.A.
Director, Division of Mathematical Sciences, National Science Foundation
(1999–2002)

Preface

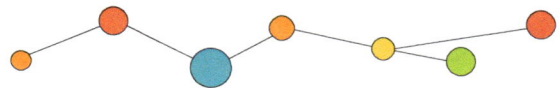

This book has been developed by the BIG Math Network, an organization supported by the mathematical sciences societies to improve networking between academia, business, industry, and government. The Network was launched at the 2016 Joint Mathematics Meetings and is supported by the American Mathematical Society (AMS), the American Statistical Association (ASA), the Institute for Operations Research and the Management Sciences (INFORMS), the Mathematical Association of America (MAA), the MathWorks Math Modeling Challenge (M3 Challenge), and the Society for Industrial and Applied Mathematics (SIAM).

In addition, the following societies have partnered with the BIG Math Network to disseminate information to their members: the American Mathematical Association of Two-Year Colleges (AMATYC), the Association for Women in Mathematics (AWM), the National Association of Mathematicians (NAM), and the Society for the Advancement of Chicanos/Hispanics and Native Americans in Science (SACNAS).

The BIG Math Network steering committee and advisory board include representatives from small and large companies and government agencies as well as representation from the mathematical sciences societies. Each of these constituencies along with our partners and colleagues have contributed perspectives to this book. We would like to offer special thanks to Philippe Tondeur, who in 2011 envisioned a program to increase the number of internships in BIG, and to Reza Malek-Madani, who raised the issue of equity with respect to access to information, internships, and jobs.

In addition, we are grateful to the following people, who shared their wisdom and experience as reviewers during the writing of this book: Sharon Arroyo, Tom Barr, Ellis Cumberbatch, Natalie Durgin, Eden Haycraft, Nick Higham, Tasha Inniss, Donna LaLonde, Lisa Miller, Whitney Moore, Nicole Morgan, Esmond Ng, Mohamed Omar, Philippe Tondeur, Suzanne Weekes, and the anonymous reviewers.

About the authors

Rachel Levy, PhD, credits her pursuit of advanced study to a senior capstone experience in operations research at the National Aeronautics and Space Administration (NASA), mentored by Bruce Pollack-Johnson. She advocates for mathematical modeling in K-16, and has supervised industry-sponsored capstone projects through the Harvey Mudd College Mathematics Department clinic program. She serves as the Vice President for Education for SIAM, and founded the BIG Math Network. Her piece on industrial mathematics in the *Princeton Companion to Applied Mathematics* was selected for inclusion in the anthology *Best Writing on Mathematics 2016*. She is the Deputy Executive Director of the Mathematical Association of America.

Richard Laugesen, PhD, collaborated with colleagues in the Department of Mathematics at the University of Illinois to create a graduate internship and training program that serves about 25 students per year, with funding from the U.S. National Science Foundation. This program offers a model for departments that want to help students make connections in BIG, yet have only a few faculty members identifying as applied mathematicians. He serves on the steering committee of the BIG Math Network, and gratefully acknowledges support from the Simons Foundation (#429422 to Richard Laugesen) and inspiration from Philippe Tondeur, former director of the Division of Mathematical Sciences at the National Science Foundation.

Fadil Santosa, PhD, has developed extensive ties to BIG through the Minnesota Center for Industrial Mathematics (1995–2007) and the Institute for Mathematics and its Applications at the University of Minnesota. From 2008 to 2017 he directed the Institute, which runs industrial postdoctoral positions, mathematical modeling in industry workshops, math-to-industry boot camps, and numerous workshops connecting mathematical scientists in academia with industry. He chairs the BIG Math Network Steering Committee, the SIAM Career Opportunities Committee, and the AMS Committee on the Profession. He is indebted to Avner Friedman for showing him the way into industrial mathematics.

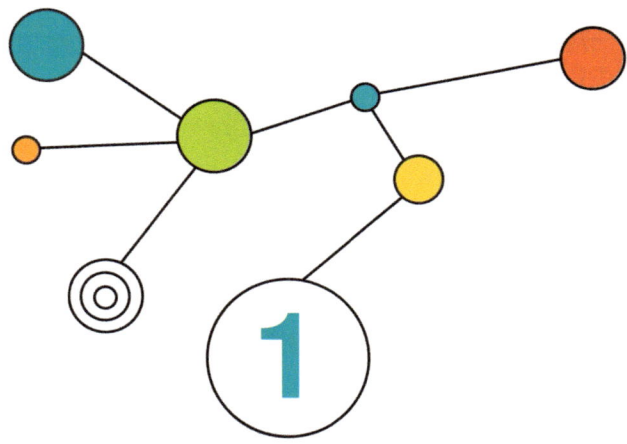

Why the *BIG Jobs Guide* is for you

Mathematical scientists are in demand more than ever. Experience in mathematics, statistics, operations research, and computational science prepares people for a host of jobs in business, industry, and government (BIG). These jobs are consistently rated among the highest areas for job satisfaction and even happiness. This book will help you become aware of interesting job opportunities and what you can do to work toward obtaining the position you want.

The terms business, industry, and government include everything from small private start-up companies, to medium and large businesses, to regional, state-level, and national governmental organizations. This includes companies that handle information (business) and manufacturing (industry), and companies that do both. In addition to business and industry, the government is a large employer for many kinds of jobs, including agencies that conduct research, provide services, and do the work of running the government itself.

People trained in the mathematical sciences have analytical and problem-solving skills that are vital in the workforce. The Bureau of Labor Statistics at the U.S. Department of Labor writes in its online *Occupational Outlook Handbook* that:

> Employment of math occupations is projected to grow 28 percent from 2016 to 2026, much faster than the average for all occupations.... Growth is anticipated as businesses and

government agencies continue to emphasize the use of big data, which math occupations can analyze.

This book helps prepare you for those careers. You can read in a non-linear fashion, taking advantage of the graphics and checklists through the book to focus on your needs.

What BIG is about

The term "BIG" as an acronym for "business, industry, and government" was coined by the BIG-SIGMAA special interest group within the Mathematical Association of America. Some people distinguish business from industry by thinking of industry as heavy industry such as manufacturing. The term BIG is useful as a positive alternative to the term "non-academic," which can have negative and even derogatory connotations. The simpler term "industry" is preferred by some, as in the Society for Industrial and Applied Mathematics (SIAM), for example.

The term BIG is not known outside the mathematical sciences, so if you walk up to a recruiter and say "I want a BIG job" they will most likely not know what you mean (or they might think you want to run the company). This book will help you learn what you *can* say to effectively communicate with recruiters about the kinds of jobs you are qualified for, and the kinds of jobs you want.

While this handbook includes general job-seeking principles, we have tailored the ideas toward people connected with mathematics, statistics, and operations research. The writers of this book, and most of the data presented, are U.S.-based. Nevertheless, much of the advice remains relevant to job seekers around the world (see quote below), and we hope that if you learn of data that exists for the U.S., you can find similar information for the country where you are seeking a job. The BIG Math Network would like to hear from readers about different career pathways in the countries where they live and work, which we could feature on our blog.

The employment landscape in mathematics in the United Kingdom shares with the U.S. all the main features that are described in the BIG Jobs Guide, and the advice that the guide contains will be of great value to students and young researchers. Over recent years in the U.K., the university sector, responding to encouragement from central government, has made great strides in raising levels of collaboration with

> *business and industry. The economic importance of mathematics to the U.K. was confirmed in an influential 2012 report from Deloitte. We are now in a period where the transforming potential of data science is rapidly expanding the scope of mathematical modeling, opening up more opportunities for new graduates than ever before. Many businesses are investing heavily in these new technologies and expanding their relationships with the academic research base. Young researchers are some of the main beneficiaries, and the BIG Jobs Guide offers concrete advice on making the most of the opportunity.*
>
> Robert Leese, PhD, Smith Institute

This book is for you

This book is intended for **job seekers at many career stages**. You may be seeking a BIG job for the first time, or thinking about changing jobs—career paths can follow interesting twists, turns, and leaps. We welcome readers who are undergraduates, graduates, postdoctoral faculty, tenure track and tenured faculty members, and people already working in business, industry, and government.

> *My 63-year-old sister Katrina has been getting A's in her computer science courses and has started a company to help people with fibromyalgia learn what affects their health through an app that uses machine learning to recognize patterns in the data they enter. It is never too late to start!*
>
> Maria Klawe, PhD, Harvey Mudd College

Maybe you are in the **middle of your undergraduate degree** program. You have time to experience different courses, research opportunities, and internships. You can build a résumé with skills and demonstrate that you are an independent learner, reliable worker, creative problem solver, and proactive employee.

Maybe you are seeking employment **at the completion of your undergraduate degree.** You want to know how to connect your experiences as a student to the requirements of various jobs, and supplement your educational experiences as necessary to get the job you want. Your financial situation might impact your flexibility when balancing income requirements against geographic preferences and career dreams. This book can help you find a job that works well for you.

Maybe you are **currently in graduate school or almost finished**. You might be in a program with strong industry connections, or a department whose members have little expertise outside academia. Some advisors might be supportive of your doing an internship or a fellowship at a national lab or company. Others could be ambivalent or even hostile to the idea, if they think it might threaten your attention to your research program or extend the time to completing your degree. You will need to navigate this territory carefully, attending to both your own needs and your obligations to your advisor and graduate program.

Perhaps you are in a **postdoctoral position**, doing research or a mix of research and teaching. Perhaps your initial thought on entering the position was that it would be a step toward a tenure track job at a college or university. Now you are not sure whether that is the path you want to take, and you would like to consider more options.

Maybe you hold a **visiting position or a lecturer position** in which teaching is your primary responsibility. You enjoy the freedom from committee work, research pressures, and advising but you are not satisfied with the length of your contract or the level of salary and benefits. Because you are not sure that you can transition from the current position to a permanent academic job, you are curious about what options might be available in industry. You do not currently have job security in academia, and so this aspect of BIG jobs is not so much of an issue. You would like to know how to position yourself to land a more satisfactory position in BIG.

Perhaps you are a **tenured faculty member** who has spent your entire career in an academic setting. This book will help you support students who seek to join the BIG workforce. Chapter 11 outlines numerous practical steps that faculty members and academic departments can take in order to mentor students toward careers in business, industry, and government.

Perhaps you work in **business, industry, and government**. You may want to hire an intern and would like to know what a person trained in the quantitative sciences can bring to the table. Maybe you want ideas about how to structure an internship. Or you are in human resources and want to build internship connections with academic institutions, departments, and prospective employees.

Maybe you **want to change your type of work** for professional or personal reasons. If you have not been on the job market for a while, this book can help you navigate the task of retraining, or reframing your skills. The jobs available today might not even have been imagined when you were in school.

We aim to promote equitable access and inclusion in our profession, broadening the notion of what it means to be a mathematical scientist, statistician, or operations researcher.

Women, we need you at the table in technical and leadership roles. Data science jobs are exciting and increasingly critical, and the societal relevance continues to grow.

Margot Gerritsen, PhD, Stanford University and founder of Women in Data Science Conference

This book is for you. Now let's get started!

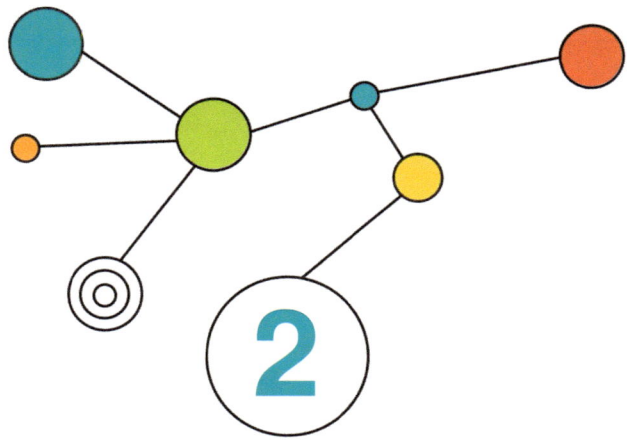

Why you need to think BIG

The job market news is generally good. Companies and government agencies recognize the value added to their enterprises by quantitative thinking and modeling. Jobs for people trained in the mathematical sciences are also becoming more visible in the media and popular culture, which raises our profile and appeal to potential employers.

This chapter begins with an overview of information sources about the BIG job market for both undergraduates and graduate students, with a focus on the U.S. situation. Some aspects might not be relevant in your local context, but the general trends are apparent in job markets around the globe. The chapter goes on to analyze the academic job market for PhD students, making the case for graduate students to explore BIG careers.

 ## BIG job market

Mathematical scientists in industry and government enjoy robust demand for their skills, and good salaries too. The *Occupational Outlook Handbook* of the Bureau of Labor Statistics at the U.S. Department of Labor says:

> Math occupations had a median annual wage of $81,750 in May 2016, which compares favorably to the median annual wage for all occupations of $37,040.

Salaries vary widely. For example, data scientists with graduate degrees in the mathematical sciences can command six-figure starting salaries. Undergraduates, especially those with computational expertise, enjoy similar opportunities to graduate students, although it can take them longer to become established in careers. For a wealth of information about working conditions and the training required for different types of positions, you can browse the website at the Bureau of Labor Statistics.

Professional societies also gather valuable data. For detailed salary reports about statisticians in BIG, you can consult the reports published every few years by the American Statistical Association. These reports are freely available at the association's website. The American Mathematical Society gathers data annually on "Employment of New PhDs." These reports are available at the society's website and include salary ranges for new PhDs starting BIG careers.

Opportunities for graduates with an associate or bachelor's degree

There are plenty of opportunities for taking a BIG job directly after your undergraduate degree. There are several reasons you might want to do so:

- You might like the idea of obtaining further education in the workplace.
- You might want to begin earning a full-time salary right away, possibly to pay off student loans.
- You might want a break from school before considering and applying for more education.

Without a higher degree it is possible that you won't obtain a managerial/leadership position as fast, and some jobs will require a higher degree to be competitive. But with the right combination of skills and experience you can position yourself for many great jobs.

Undergraduates in statistics and operations research usually get solid preparation for BIG jobs. Their degrees are easy to "sell" because employers know what a statistician or an operations researcher does. An undergraduate degree in mathematics is different. It takes some work to show potential employers the value of your training. No matter what your major, go to your college's career center and find out where the graduates in your major are going. You'll find that many have found positions in BIG. Admittedly, some may not use much of their training on a daily basis, but it is worth pointing out that once you are inside a company, you can often navigate your way into the kind of work that you like to do.

Academic job market trends

This section is for PhD students and postdoctoral faculty in the mathematical sciences who would like to consider an academic career. From a research perspective, academics are very similar to the self-employed. They are typically judged, and promoted, as individuals rather than team members.

If your preferred career is not in academia then you may wish to skip this section and move on to the next one. If you *do* want to want to work in academia, then keep reading, because it's best to go in with your eyes open. In this section we will answer the following questions:

- What are the trends in PhD production nationwide?
- How many postdoctoral positions are available?
- How many tenure track positions are filled each year?
- What percentage of PhD graduates end up in academic positions immediately after graduation?

How many PhDs are produced each year?

Mathematical sciences PhD production in the U.S. has almost **doubled in the past 15 years.** During the 1990s, around 1100 PhDs were awarded annually. That number dropped in 2001–2002 to 960, and then almost doubled to the most recent total of 1901 graduates.

Figure 2.1 shows the trend. The increase is most pronounced in statistics and biostatistics (up more than 150%) but is also massive in mathematics (up about 80%).

How many new PhDs get a postdoctoral position?

A postdoctoral position is essentially required for any mathematician pursuing an academic research career. The same increasingly holds for the upper tier of teaching-oriented positions at liberal arts colleges.

In biostatistics, postdoctoral positions are almost a necessary first step because, in general, biostatistics faculty are required to generate substantial grant support for their positions. Postdoctoral experience is less standard for statistics PhDs because the job demand is greater than supply.

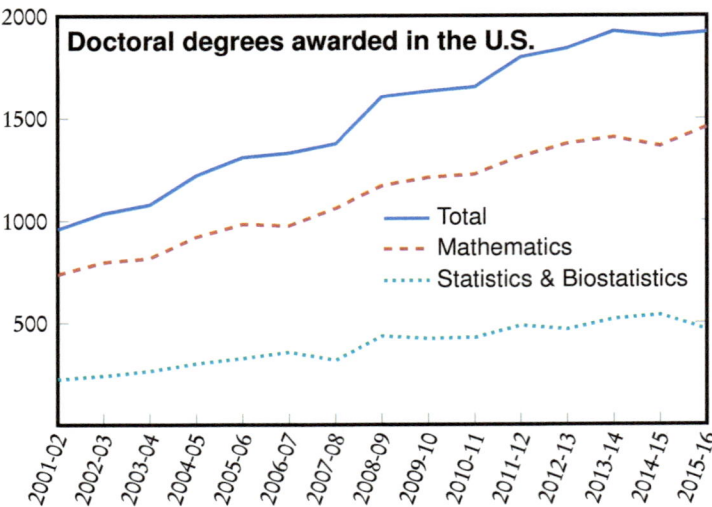

Figure 2.1. PhD production in mathematics, statistics, and biostatistics in the U.S. has increased dramatically. Source: Figure A.2 in *Report on the 2015–2016 new doctoral recipients*, Notices of the American Mathematical Society 65(3) (2018), 350–364.

Less than a third of new PhD graduates go into postdoctoral positions: Figure 2.2 breaks down the numbers by type of degree-granting department. Overall, 59% of new mathematics PhDs from large private universities got postdoctoral positions in 2015, 45% from mathematics departments at large public universities, 28% from applied mathematics, and 19% from statistics/biostatistics groups. Further, getting a postdoctoral position does not guarantee a subsequent tenure track job. Many postdoctoral faculty members who envisioned an academic career for themselves must make the shift to industry or government at that stage.

Many new PhDs who worked hard and wrote strong dissertations will not find postdoctoral or faculty positions in the U.S.

How many tenure track jobs are there?

Around 800 tenure track or tenured positions are filled annually in the mathematical sciences, according to the AMS *Report on 2015–2016 academic recruitment, hiring, and attrition* (Notices of the American

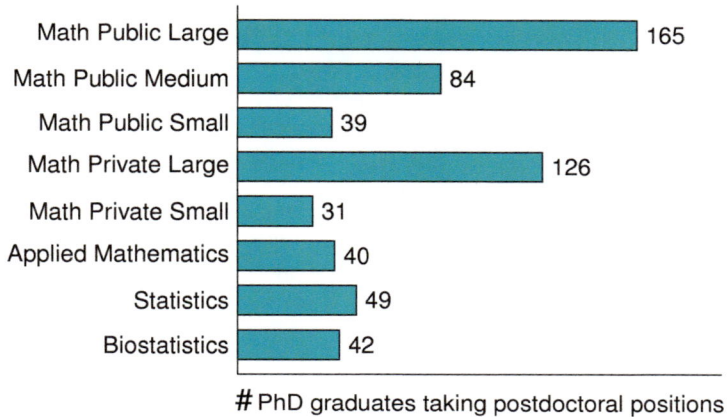

PhD graduates taking postdoctoral positions

Figure 2.2. Number of doctoral graduates in the mathematical sciences who go into postdoctoral positions, from various types of doctoral institution. Overall, less than a third of new PhDs go into postdoctoral positions (roughly 600 out of 1900 PhDs, in 2016). The postdoctoral rate varies by type of degree-granting institution: 59% of new mathematics PhDs from large private universities, 45% from mathematics at large public universities, 28% from applied mathematics, and 19% from statistics/biostatistics groups. SOURCE: Figures A.1 and E.4 in *Report on the 2015–2016 new doctoral recipients*, Notices of the American Mathematical Society 65(3) (2018), 350–364. That data on applied mathematics, statistics, and biostatistics is not broken down into public and private institutions.

Mathematical Society 64 (2017), 584–588). That total is a lot less than the number of PhDs produced each year. The academic picture is brighter in the subfields of statistics and biostatistics, where about 100 postdoctoral positions are available (according to Figure 2.2) and roughly the same number of tenure track or tenured positions are filled.

To illustrate further that the number of academic openings is substantially less than the number of PhD graduates each year, Table 2.3 presents a breakdown of the academic openings advertised recently in North America. Approximately 450 tenured or tenure track positions were advertised, which is a far smaller number than the approximately 1900 PhD graduates in the U.S. annually.

For mathematical scientists who get one of these tenure track positions, we suggest some caution about joint appointments, sometimes called interdisciplinary or cross-department appointments. These can look good on paper, but issues that must be addressed at the hiring stage include: different requirements and measures of success in different

Position type	#Ads
Tenured/tenure track faculty	452
Non-tenure track faculty	124
Postdoctoral position	167
Fellowship or award	10
Administration	4
Academic admissions	4
Student programs	2
Other	37+

Table 2.3. Snapshot of academic positions advertised on Mathjobs.org, the main academic job site in North America, in November 2017. The number of academic openings is substantially smaller than the roughly 1900 PhD graduates in the U.S. each year.

departments; lack of a single mentor who cares about and understands your whole career; and overload because each department makes heavy demands on your time. It is best if a "home" department is specified.

How many new PhDs take a job in academia?

Just over half of new PhDs start out employed in some kind of U.S. academic position—see Figure 2.5. This total includes postdoctoral and tenure track positions, non-tenure track and "visiting" positions, and academic employment outside the mathematical sciences. The situation is changing rapidly for new PhDs. The U.S. academic employment rate has dropped noticeably in the past five years.

What about jobs at two-year colleges?

The student population in two-year colleges is exploding. According to the April 2016 College Board Research Brief, "In fall 2014, 42% of all and 25% of full-time undergraduate students were enrolled in community colleges." Jobs in two-year colleges can be attractive because they can have the security of the tenure track coupled with the advantages of a teaching position. Since these programs tend to be relatively affordable and don't require housing or meal programs, they serve a segment of the population that can get a significant economic boost from higher education. These colleges also provide a bridge for adults returning to school and students who need to build a successful academic profile before transferring to a four-year school.

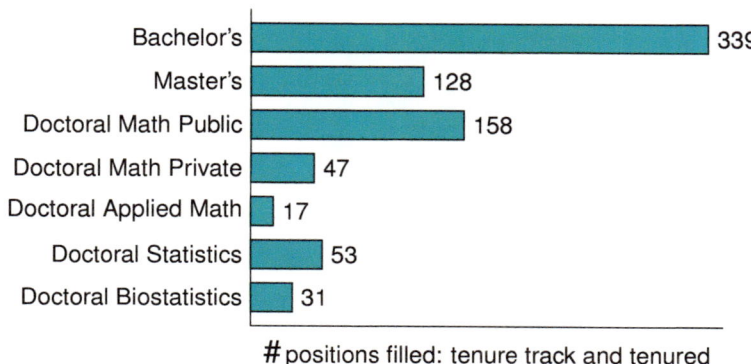

positions filled: tenure track and tenured

Figure 2.4. Bachelor's and master's level institutions accounted for about 60% of tenure track and tenured hiring in 2015–2016. Further, the total number of tenure track and tenured positions filled (750–800) amounts to only about 40% of the annual PhD production (1900). SOURCE: Figure F.2 in *Report on 2015–2016 academic recruitment, hiring, and attrition*, Notices of the American Mathematical Society 64(6) (2017), 584–588. The data there on master's and bachelor's level institutions is not divided into categories of public/private, applied mathematics, statistics, or biostatistics.

To teach at a community college you must have an advanced degree and expertise in a subject needed at the college, as explained in *How the Job Search Differs at Community Colleges* by Rob Jenkins (Chronicle of Higher Education, 2013). He writes that the interview process is different too:

> You will probably be one of 10 or 12 candidates, perhaps for more than one position, with as many as half of those candidates being interviewed by the same committee on the same day. You will meet with committee members for an hour, maybe 90 minutes, during which you will answer questions and give a short teaching demonstration. Afterward you may meet with a department chair or dean, but probably not. The community college may or may not cover your travel expenses (something to ask about when the invitation is extended) and will almost certainly not take you to dinner.

Make sure you can provide detailed examples of teaching experiences (both successful and challenging) that you can discuss in detail.

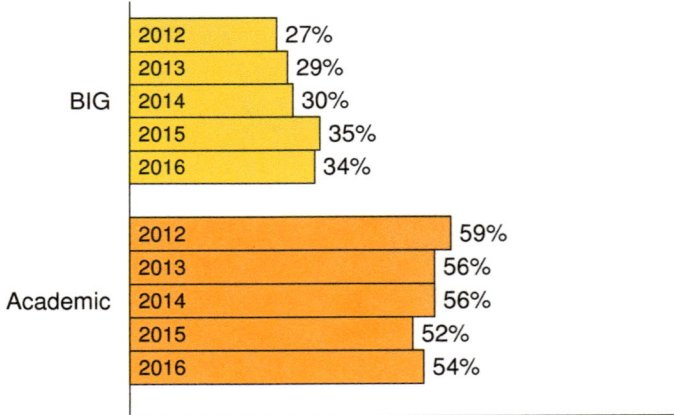

Trend in first jobs of U.S. PhD graduates

Figure 2.5. More PhD graduates than previously are going into business, industry, and government (BIG), while the percentage taking academic positions in the U.S. has declined. About 12–15% of PhDs take positions outside the U.S. (not shown), mostly in academic roles. DATA SOURCE: Figure E.7 in *Report on the 2015–2016 new doctoral recipients*, Notices of the American Mathematical Society 65(3) (2018), 350–364.

What about careers in teaching-oriented colleges?

Many undergraduate-focused institutions will hire a new PhD directly into a tenure track position without requiring postdoctoral experience. These institutions are often small, though, and the number of such openings is therefore limited.

Teaching-only positions in research departments are sometimes tenure track (although usually not). These teaching tenure track positions are often called Professor of the Practice or Professor of Teaching, and tend to carry heavier teaching loads than positions that require research. If you pursue one of these positions, check how it is positioned within the department. For example, would you have full voting rights in the department and at faculty meetings? Would you follow the same sabbatical schedule? If you get external funding, will you be allowed to use those funds for a course release? Think ahead of time about what you want from this type of position.

Non-tenure track positions are another option, with titles such as Instructor, Lecturer, Academic Professional, Teaching Professor, or Professor of the Practice. These positions prove satisfying for some people because they usually do not require research and have limited service requirements. They may lack the job security and academic freedom provided by tenure. Salaries are generally substantially lower than on the tenure track, and non-tenure track faculty can find they are shut out of departmental decision-making.

These adjunct positions may not always provide a stable career. These jobs can, however, provide enjoyable teaching experiences for someone seeking just a temporary stint. **If you are contemplating an adjunct position in which you are paid a flat rate for each course taught, then we strongly advise you to read this book and take action to investigate a rewarding career in business, industry, or government.**

What does this mean for you?

Most PhD graduates in the mathematical sciences in the U.S. will spend most of their career outside academia. Those who do remain in academia will find the majority of tenure track and tenured positions are at teaching-oriented master's and bachelor's level institutions. Non-tenure track faculty numbers are growing at all types of institution, but these positions are not always desirable.

Opportunities in BIG are expanding in both availability and attractiveness. To explore these careers, read on to the next chapter!

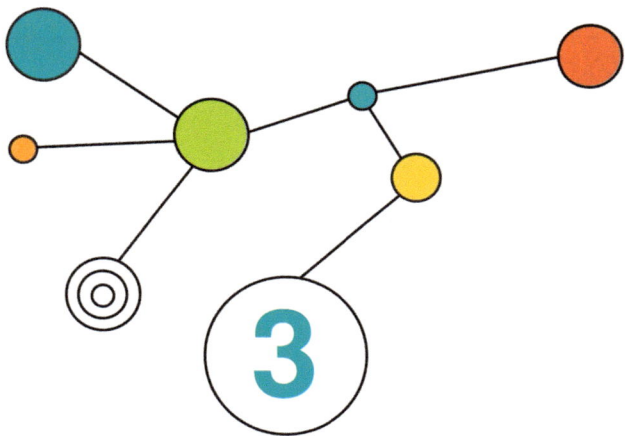

Who you are: your "special sauce"

The biggest things that a mathematician brings are the ability to take a problem and express it in a way that non-experts understand, and the ability to figure out how to best solve it. I think that our mathematics education gives us very general problem-solving skills. I don't have to use standard recipes for how to solve particular types of problems. I can pull in different ideas to create a new recipe and I have an overall understanding of why one method might work when another doesn't.

Genetha Gray, PhD, People Analytics, Salesforce

The first question: who are you?

Before you start researching jobs or attending career fairs, you need to answer the question, "Who am I?" Knowing who you are will help you not only seek the kinds of jobs that will work best for you, but will also help you communicate what you can bring to potential employers. **Values and life priorities are crucial to address at the start of the job search.** The job search should not be about what you **can** do, but what you **want** to do.

If you are a vegetarian, you do not want to work at a meat-packing plant. That is just not going to match up and you are not going to be able to do that for very long. Just like you figured out what school to go to by determining where you want the school to be, what you wanted your class size to be and what are those things about the school that are important to you. It works the same way when you are finding a job. Figure out your values, and use that to figure out what kind of organization is right for you.

Nicole Morgan, SPHR, SCP

When you work in BIG, you will need to do work that is of value to your employer. You will likely be called on to do work that was prioritized by people higher up in the company. Not every great idea you bring to the table will fit into the critical paths the company has selected to pursue.

Below is a list of questions you can ask yourself as you start identifying what is important to you and define your work values. If you find the list helpful, remember there are many online self-assessment tools and interest inventories that can help further refine your values, goals, and aspirations. In the words of Eden Haycraft at the University of Illinois career center, an assessment tool can help you "step away from the idea of a linear connection between your major and your career." She further observes that "You can't look for something you don't know you want!"

> **Use this questionnaire to begin answering the question "Who am I?" for your résumé**
>
> - Is there something that makes you want to go to work each day? What brings you joy?
> - What size organization do you want to join? How established? Some companies have been steady employers for generations, while others are just getting started.
> - What kind of working environment do you want? How informal, how noisy or quiet? Do you want coworkers who only work together or ones who also hang out during off hours? How many hours do you want to work each week, and at what times? What kinds of diversity do you seek?
> - How much job security do you want and how much risk are you willing to take?

- What salary and benefits are you hoping the job will provide? Think broadly about items such as health care, childcare, elder care, vacation, stock options, and retirement.
- Where in the country do you want to work? Do you want to work remotely? Do you want to travel for work?
- What kinds of problems do you want to solve? Are there particular tools that you enjoy using?
- Do you want to be a trailblazer? Or to mentor and collaborate with coworkers who have similar skills?
- What kinds of jobs might pose ethical conflicts for you? Are there products and services you especially do or do not want to help provide?
- Do you want to work on one project deeply for a long time or move from project to project?
- How flexible are you when someone asks you to stop what you are doing and focus on something else?
- Where do you want to find yourself in a decade? Do you want to become a manager? What professional development opportunities do you want to take advantage of?
- What do you hope to contribute to your organization and society?

Creating and sharing your work identity

Recruiters are looking for your authentic self, or what senior certified human resources professional Nicole Morgan, SPHR, SCP calls your **special sauce**—all the things that make you uniquely who you are. Once you have a sense of who you are, you can find ways to communicate this information in your résumé. You can also share your special sauce by uploading project reports, publications, and programming projects to an online repository. This shows potential employers evidence of your taking the initiative and demonstrating skills.

Some college campuses have online job boards where you can enter keywords, upload your résumé, and register for weekly updates. Make use also of the career websites listed in Chapter 6. Each of these

Recruiter by Rachel Levy

venues gives you the opportunity to explain who you are—to share your special combination of abilities, interests, and passions.

Once you know what kind of jobs you want, you can ask the director or advisor in your program for suggestions of industry contacts and alumni to talk with. You can also ask the campus career center for suggestions of contacts and companies that could help you get started in your search, and you can go to career fairs. By thinking broadly about fields that interest you, and then carrying out informational interviews with people in those fields, you can decide whether those jobs would fit with your values.

So how do you go about sharing your special sauce in a way that moves the job search forward? Here is what you don't want happening at a career fair (based on a true story)....

> **You (walking up to recruiting table):** Hi! Does your company hire math majors?
> **Recruiter (dismissively):** No, sorry.
> **You (about to give up):** You don't hire anyone who works with data or does forecasting and modeling?
> **Recruiter:** Oh sure. We have several groups with people doing that kind of work.
> **You:** Great! Could you put me in touch with the recruiter who hires for those groups?
> **Recruiter (sheepishly):** Uh, that would be me.
> **You (now worried you have embarrassed or annoyed the recruiter):** OK, well, may I have your card?
>
> Unsure what to do next, you walk away with the card.

And here is how it could go much better, if you know who you are and how to describe your special sauce....

> **You (walking up to recruiting table):** Hi! I am interested in working with data. Are any of your groups hiring in a related area?
> **Recruiter:** Yes, can you tell me a little about your background?
> **You:** I will graduate in May with a degree in mathematics, which includes course work in statistics, computer science, and data science. I especially enjoy forecasting and modeling, and I'd like to work for a start-up. I am comfortable working in R and MATLAB, and sometimes program in Python or Mathematica. I have some familiarity with Java.
> **Recruiter:** Oh sure. We have several groups with people who do that kind of work.
> **You:** Wonderful! What would be the next steps to connect with managers who hire for those groups?
> **Recruiter:** I'll do an intake interview now and you can follow up by submitting some materials online. The next step would be a phone interview and then possibly an on-site interview.
> **You:** Great, I'm available this afternoon—do you have a spot available?

These scenarios hint at the importance of the **elevator pitch**. At all times, you should have a few sentences ready to go that quickly give a sense of who you are and what you are aiming for professionally. This pitch must be brief, so that you can deploy it when you find yourself riding an elevator (or a lift, for readers in Britain) with a person who could advance your career.

So create a one-minute spiel about yourself. Practice it on friends and family, and especially people who are not in your area of expertise. Ask for their honest feedback. Keep revising, improving, and practicing your pitch.

Reprinted with permission from Randall Munroe, XKCD comic #1364.

Then later, when you have just a brief moment to make a strong impression on a recruiter or an executive at the company you want to work for, you will feel confident and well prepared.

In career fair and social situations, your elevator pitch can help you network and find people who know about jobs that are right for you. When you find potential mentors, use the elevator pitch to tell them you are interested in careers in their industry, and find out what connections and advice they can offer. Further down the line, if you start your own company then you'll need to make a pitch in front of investors and your hours of practice will certainly come in handy.

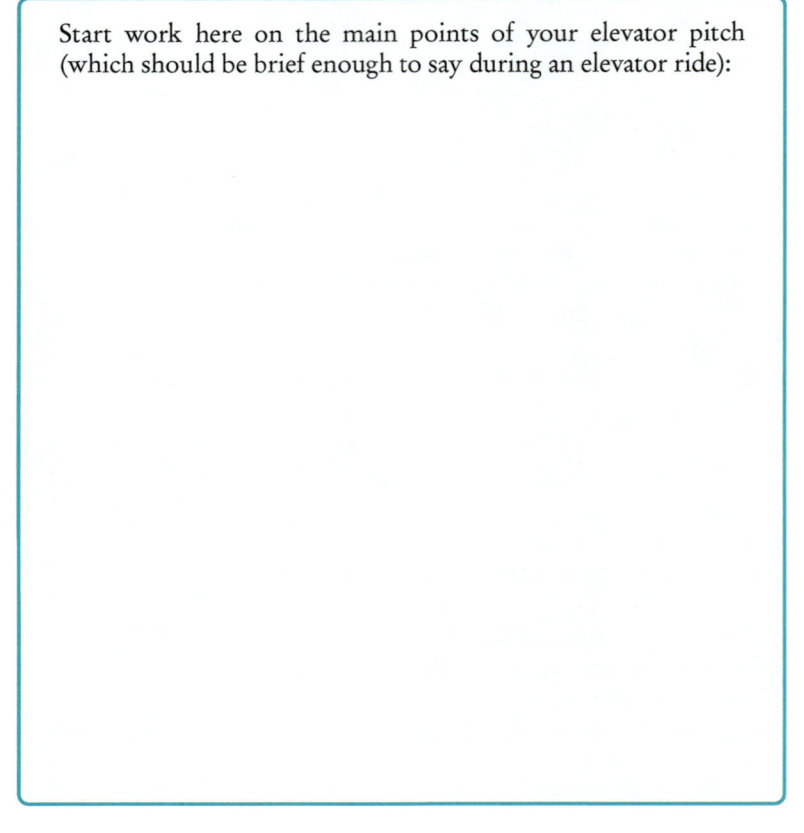

> Start work here on the main points of your elevator pitch (which should be brief enough to say during an elevator ride):

An elevator pitch is most effective when it ends with a "call to action." At the end of the well-crafted pitch the person hearing the pitch will be engaged when you ask them a question that elicits a response. It can be as simple as, "Do you know anyone who works at XYZ Communications?" or "Who do you know who needs help with this type of problem-solving?" Initiating a two-way conversation with someone is more likely to provide insights and make connections.

What employers might not know about you

Employers are looking for someone with sophisticated quantitative and problem-solving skills who can:

- figure out what questions to ask,
- examine data critically in order to answer those questions,
- determine what questions cannot be answered with the available data,
- communicate results of an analysis clearly, concisely, with focus on the "bottom line" for improvement.

Employers may not have a clear sense of the difference in training and skills between students with majors in mathematics, statistics, or operations research. Employers often do like to see mathematics plus something else—say, mathematics/statistics/operations research plus computing, or mathematics/statistics/operations research plus liberal arts.

Job applicants need to be able to describe how their skills and knowledge could be applied in practice and for a particular industry. As career counselor Eden Haycraft remarked: "Employers focus more on skills than on majors." And because most degree programs in mathematical sciences are not focused on a career path to industry, you might need practice translating your abilities into terms that resonate with people in industry.

An undergraduate degree in **mathematics** is an asset. You know the popular belief that mathematics is hard. The fact you've made it this far already means to many people that you are smart. However, if they are the employer, as much as they like to hire smart people, they may not know what variety of things you can do for them. It is your task to educate your potential employer about who you are and what you can do.

Statistics and **operations research** degrees are better understood. BIG organizations hire many statisticians. The fact that they are present in large numbers means there is an understanding of the value of statistical training. The same goes for an operations research degree. Employers more or less know that your training means that you can translate business problems into mathematical models, which can be used to find optimal or near-optimal solutions. These problems may include scheduling (e.g., airline crews), transportation (e.g., shipping of products), assignment (e.g., workers to shifts), inventory, facility location, and so on. They have a general knowledge of what they are getting when they hire you. They know you are skilled in problem-solving.

It is your responsibility as the job seeker to relate your experiences to the job you are seeking. This translation of your skills into recognizable terms helps to communicate your abilities and potential. For example, if you want to work in signal processing, you may not have had a course in that topic, but you can explain that your strengths in linear algebra and computational tools are directly applicable to those kinds of problems, and you can point to your ability to analyze data and meet deadlines.

Graduate students can communicate about the **process** of doing research rather than the specific content of their thesis, because process skills such as scoping, planning, and executing a project are highly transferable.

So as you think about who you are, and where you stand out from the crowd, remember you do have special talents and abilities!

> *Not every mathematical terminology or theorem will be used when you work in industry, but there is one place where mathematical training helps you stand out: the capability of abstraction and thinking fundamentally. It will always lead you to new solutions.*
>
> Dongning Wang, PhD, American Family Insurance

Chapter 3 checklist

☐ Think through answers to questionnaire.

☐ Develop your elevator pitch with a specific audience in mind.

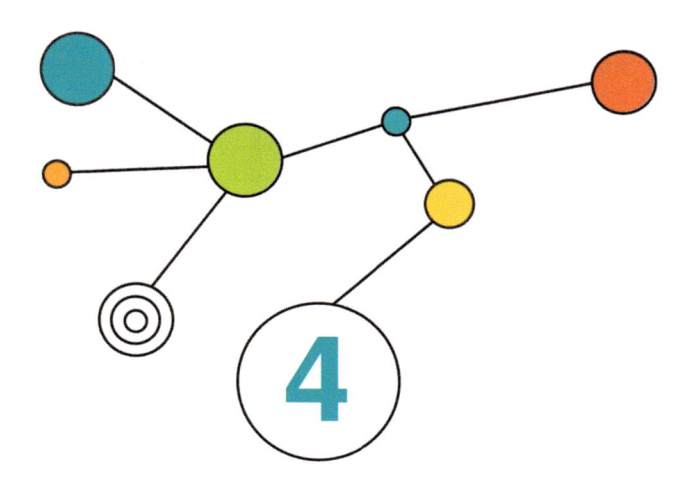

What you should study

I wish I had known in college all of the opportunities available to someone with an advanced degree in mathematics. With a doctorate in applied mathematics, I have had the opportunity to work in academic institutions as a tenured faculty member, to work in the federal government in the Directorate for Education and Human Resources at the National Science Foundation, and now in a non-profit professional society, INFORMS. I have been able to use my skills and training in each of these settings. Mathematics is ubiquitous! Being trained as a mathematician means you are a problem-solver because you know how to think logically and critically. For me, with my background in operations research, I am always seeking ways to make processes more efficient.

Universities are in a remarkable position to expose students to the wealth of career pathways for mathematicians. The options are almost endless!

Tasha Inniss, INFORMS, PhD

You probably feel you don't have enough useful skills to get a job in BIG. We believe you do. Remind yourself to think positively about the skills you already have, be realistic about where your preparation can be strengthened, and initiate action **today** to acquire additional skills—preferably right after reading this chapter.

Becoming a strong candidate

To become a strong candidate for internships and jobs, your portfolio should include evidence of the following:

- foundational courses in mathematics and statistics;
- specialized courses in your discipline;
- computing course(s) in a modern language, e.g., C++, Python;
- experience using software packages such as R, MATLAB, COMSOL;
- communication skills;
- completion of projects;
- written and oral reports, both "in progress" and "final";
- a course with substantive team projects is desirable;
- basic knowledge and experience in data science and analytics is optional, although certainly desirable these days.

Whatever you put in your portfolio, make sure that you can concisely describe the goals, methods, and accomplishments of the project, including any challenges you overcame and any skills you picked up or methods you developed to solve the problem. If you cannot outline your own contributions to the results, do not list the project.

This chapter contains a list of courses you should consider taking and the reasons why BIG employers deem them valuable. Some of the material can be acquired through independent reading, either with a professor, a group of students, or by yourself. If you decide to go it alone, remember to look online for resources to help when you get stuck.

As you are solving problems in your courses, apply a forward-looking mind-set. Ask yourself, how can this concept, algorithm, or theorem be implemented in practice? Is the problem solvable on my laptop if it has 1 million variables? Would an approximate solution be good enough in practice, and if so, how quickly can it be found?

The list of courses forms a "wish list," not a set of requirements. Not all of the courses are needed for every career direction. In other words, regard the list as an appealing menu from which you will choose some tasty and nutritious dishes. After getting a few of the courses under your belt, you will be ready for meaningful conversations with potential employers.

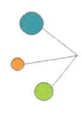

Foundational courses

Your training in the mathematical sciences provides a powerful toolkit for tackling all kinds of real-world problems. From subjects such as trigonometry, calculus, linear algebra, linear programming, probability, statistical regression, topology, combinatorics, and numerical analysis you will gain conceptual frameworks with which to analyze information, build a model, and make quantitative predictions. Here are some foundational courses in the mathematical sciences you should master.

Differential equations

Differential equations play a significant role in modeling physical, chemical, biological, and engineering processes. It is even better if you can pair your theoretical courses with courses that teach computational solution methods and software tools.

Discrete mathematics

Mathematics, computer science, and operations research departments all usually offer discrete mathematics courses, which can be application driven, theorem-and-proof oriented, focused on algorithms, or any combination of the above.

Discrete mathematics involves a different kind of thinking from that used in continuous mathematics courses such as a course on differential equations. Methods include combinatorial analyses of counting, enumeration, and optimization problems, and applications to problems in graph theory. Tools in this area include probability and logic.

Linear programming

Linear programming refers to the process of maximizing or minimizing a linear function subject to linear and positivity constraints. This powerful operations research technique is used throughout industry, from scheduling to online advertising to supply chain management. Mixed integer linear programming/optimization is also used a lot.

The process of transforming a business problem into its linear programming description is often done through a "modeling tool" such as AMPL, which takes a problem description and its parameters and produces the cost function and constraints needed by the linear programming solver. In applications, some variables may be continuous while others take on discrete values.

Linear algebra

Linear algebra provides a framework for understanding relationships that may be complicated but are nonetheless linear. Linearity is ubiquitous in applications, because the first step in solving many nonlinear problems is to linearize.

You will grasp foundational concepts such as bases, dimensions, positivity, subspaces, orthogonal complements, diagonalization, least-squares, and factorizations such as the singular value decomposition.

On the practical side, you will learn how to solve linear algebra problems on the computer (with MATLAB, Maple, Mathematica, Python, R, or other software packages). These courses can also teach you how to compute problems at scale; many industrial problems involve matrices with millions or billions of rows and columns, and so it helps to learn iterative methods and approximate methods. In industry, a good approximate solution is often good enough!

If possible, take two semesters of linear algebra: an introductory course and then an applied or computational course. There is no such thing as knowing too much linear algebra.

Mathematical modeling

Many courses provide students with exposure to some models, such as a mass–spring–damper system in a differential equations course, but in BIG jobs it is more likely that you will need to create your own models or explain why one does or doesn't work for a particular situation. To get a taste of the ways mathematical modeling plays a role in the workplace, find a modeling course in which you tackle problems that are open at the beginning (as you focus the problem and make assumptions), the middle (as you choose mathematical and computational tools to formulate your model), and at the end (as you test your results, iterate to improve, and then communicate the best solution you can devise in the time allotted). You may learn about simulation, a technique to inform modeling. If no such course is available, you can join modeling competitions (such as the MCM/ICM), industrial mathematics workshops, or datathons to get experience.

Numerical analysis/scientific computing

An introductory course in numerical analysis and scientific computing will likely expose you to computational linear algebra. Courses at the advanced undergraduate level go a lot deeper. You will study quadrature methods for integration, approximation of functions, and interpolation. Most likely such a course will include basic methods for solving a

$R^2 = 0.06$ REXTHOR, THE DOG-BEARER

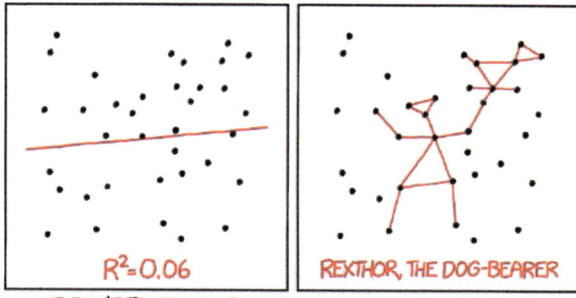

I DON'T TRUST LINEAR REGRESSIONS WHEN IT'S HARDER TO GUESS THE DIRECTION OF THE CORRELATION FROM THE SCATTER PLOT THAN TO FIND NEW CONSTELLATIONS ON IT.

Reprinted with permission from Randall Munroe, XKCD comic #1725.

system of nonlinear equations, ordinary differential equations, and even partial differential equations. An important aspect of such a course is how to demonstrate error analysis and convergence. Such knowledge is crucial when you solve a problem and must describe the accuracy and stability of your method and solution. For example, your code might forecast the weather one month in advance, but is that prediction accurate?

Probability

Most real-life problems are fraught with uncertainty, and probability is therefore an important tool for mathematical scientists. Probability allows you to give "on average" and "worst-case" answers, and to quantify uncertainty. It also provides the foundation for statistics. A first course in probability requires little more than calculus as background. At the graduate level, probability relies on real analysis (measure theory). Counting problems that arise in dice throws and card games are surprisingly useful in real-life situations, and so it is no surprise that interview questions asked by financial firms are often probabilistic in nature.

Statistics

Statistics is of central importance in BIG jobs. Most projects will involve data, and if you want to analyze data, either big or small, you'll need statistics. Statistics courses can focus on either theory or practice, so choose your path carefully. You should come away with a firm understanding of hypothesis testing, regression, estimation, and methods such as analysis of variance and principal component analysis. If you can, find a course that introduces you to time series and tools associated with their analysis and modeling. Many statistics classes now incorporate the open source programming language R, which is increasingly used in BIG jobs.

> *I wish I had studied applied mathematics. Either that or a separate major in statistics and machine learning.*
>
> Adil Ali, PhD, Wells Fargo

Stochastic processes

A course in stochastic processes is desirable although not necessary. Random walks, Markov chains, branching processes, martingales, queuing theory, and Brownian motion all have real-life applications. These concepts, as well as Wiener process, stationary sequences, and Ornstein–Uhlenbeck processes, find applications in many areas of business and industry that rely on modeling and predictions.

Undergraduate capstone course

Does your undergraduate major include a capstone activity? It might be a senior project or senior thesis, or a team project. Engaging in this kind of activity shows that you can dig deep and stick to something. It also provides substantive material to talk about in an interview. The nature of your capstone project need not relate directly to what you plan to do next in your job.

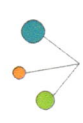

Disciplinary courses

Following is a list of courses that might be offered at your institution, by various departments. Remember the list is a menu—if you have the opportunity and the curiosity, take a few of these courses and you'll be surprised by how much they build on your mathematics, statistics, or operations research background.

These courses are rich in examples of quantitative thinking used to solve real-world problems. Many of these courses also provide opportunities to hone your computing skills. Courses may include projects that can be done alone or in a team, thus adding to your experience and portfolio.

Whether you are in mathematics, statistics, or operations research, we recommend you take courses in the other departments to strengthen your skill set and foundational knowledge. If you are an undergraduate, think about doing a double major with computer science. If you are a graduate student, your program might require (or at any rate encourage or allow) courses outside your discipline. If you plan carefully, you might be able to complete a master's degree in another discipline with a little additional effort. For example, many PhD students acquire a master's degree in computer science simply by taking a few courses outside their department.

Reprinted with permission from Randall Munroe, XKCD comic #1838.

Coding theory and cryptography Computer security and privacy are important areas in all industry sectors. Coding is essential for communications and data storage. So acquiring basic knowledge about these two areas is definitely a plus. Moreover, with cryptocurrencies and blockchains rising in importance, gaining mathematical knowledge about this topic is an asset.

Computer vision This is usually offered through computer science or electrical engineering departments. It can be heavily mathematical, using ideas from Fourier analysis, differential geometry, and other areas of mathematics to solve the problems. This type of course might also be called image processing or remote sensing.

Continuum mechanics Physics-based models continue to play an important role in applications. For example, modeling thin films of fluid flow on a surface can be important in the manufacturing of coating processes, and for treating lung problems in premature babies. To develop accurate models for this purpose, you need continuum mechanics and fluid mechanics. These courses can be highly mathematical, involving partial differential equations, and are usually offered through mathematics, physics, mechanical engineering, or aerospace engineering departments.

Design of experiments Usually offered through the statistics or biostatistics departments, it is central to areas such as clinical drug trials. These courses are critical to anyone planning to conduct applied research using statistics, to ensure that studies are conducted in a well-designed and scientifically valid manner. You can imagine here many different types of experiments, such as laboratory, algorithmic, modeling, or simulation.

Electromagnetics Light and radio signals propagate as electromagnetic waves. If you want to end up in an electronics or nanotechnology company, you should consider taking a course in electromagnetics that covers both the theory and applications of Maxwell's partial differential equations, which provide the fundamental model in the area.

Graph theory, combinatorics, network optimization These topics are usually offered through mathematics departments, or in theoretical courses in computer science. Graph theory is important for modeling any kind of network, such as communication networks, social networks, and the power grid. Combinatorics more generally, especially combinatorial optimization, shows up in many business and engineering problems. Combinatorial optimization problems tend to be difficult or impossible to solve exactly, and much effort goes into developing powerful approximate methods.

Introduction to operations research Such a course is usually offered through industrial engineering or operations research departments. The course will introduce ways that business problems are analyzed and modeled using mathematical language. They often end up being probabilistic models whose behavior can be described or optimization problems that can be solved with existing tools.

Machine learning/data science/artificial intelligence Many of the jobs available to mathematical scientists have something to do with machine learning, data science, and artificial intelligence. They all use computational techniques to analyze data for the purpose of learning, predicting, and decision-making. We recommend that you take a course in these growing areas.

Data science programs are popping up in many universities but the field is not yet well defined, and data science courses on different campuses can therefore cover completely different material.

We advise you to find out what the course plans to cover. Roughly, you can expect statistical methods for data analysis, neural networks, and other machine learning techniques for classification, clustering, and data modeling.

If you cannot find a course on your campus that suits your needs, then seek a good online course. (See the advice later in this chapter.) Read the reviews and, if possible, get a recommendation from someone who has completed the course. Many students start these courses and do not complete them. You will want a high-quality course to learn from, and finish.

Ordinary and partial differential equations   Ordinary and partial differential equations remain important because many complex physical processes are modeled by them in industrial research and development, whether it is by car companies or medical device makers. A theoretical understanding of these topics is important, even though in practice you'll be dealing with computer model approximations and numerical solutions. You need the theory in order to understand the simulations and their reliability (or unreliability). For instance, it helps to know before you do any computations whether the solution is expected to blow up in finite time. Also, how sensitive is the solution to the choice of initial conditions and parameter values?

Signal and image processing   Signal and image processing shows up in automation. For example, by using image processing, a company that does circuit inspection might be able to automate detection of flaws in circuit boards. Signal processing is used to enhance and improve the audio signals in hearing aids. This is a mathematically rich area.

System simulation in operations research   When systems are too complicated to be analyzed, computer simulation becomes an important tool to understand instances of behaviors or to attempt to understand average behaviors. The field of agent-based simulation is a new area, where, for example, one can model traffic in a realistic way. Such courses are normally offered in operations research departments, although some computer science departments offer similar material.

Time series analysis   This area of statistics is dedicated to the study of sequential data records over time. Examples include financial data (stock prices, interest rates, trade volumes), store sales data (number of widgets sold per day), weather data (temperature, rain, wind), and medical data (blood pressure, temperature, pulse). This is a mature area with many tools for analysis that can provide useful insights for decision makers.

Of course we could go on suggesting courses. If you have exhausted this list, there are plenty of other terrific course options, such as data structures, distributed computing, robotics, control theory, and optics. There's not time to do everything, so enjoy picking and choosing courses that interest you the most!

Summer programs

You can get a lot out of summer research experiences, whether they are a Research Experience for Undergraduates (REU) at an academic institution, a research experience at a national lab, or a research position within a company. You can find many of these opportunities on a site called MathPrograms.org, which also provides a way to apply and upload recommendation letters. Official REUs are funded by the National Science Foundation; in addition, there are summer research experiences and workshops for both undergraduates and graduate students in many locations and institutions. The best way to find these opportunities is to search online. You might also learn about them through mentoring sites hosted on social media or by the professional societies.

Competitions

If you enjoy taking part in competitions—perhaps you did the M3 Mathworks Math Modeling Challenge in high school—then you could form a team and enter one of these contests for college students:

- the Mathematical Contest in Modeling (MCM) and the Interdisciplinary Contest in Modeling (ICM),
- the ASA Datafest,
- the Kaggle challenges in data science,
- hackathons and coding meet-ups, and
- the INFORMS Operations Research & Analytics Student Team Competition.

Online courses

The major online vendors offer many high-quality free courses. You will find good courses at sites such as:

- Coursera,
- EdX,
- MIT Open Courseware,
- Udacity.

Searching for courses Online offerings change rapidly, and so it is best if you identify **high-quality online courses** for yourself. Here are our search tips:

- Start by searching for "free online courses" and browse the results to get a sense of the major vendors and what they each emphasize.
- Narrow the search to a particular subject area of interest, e.g., "free online course regression."
- Find knowledgeable reviews, so that you can select a good course.

Check the reviews User reviews are a starting point, although they can reflect the level of the student more than the quality of the course. It is helpful instead to find a comparative review of course offerings, written by an experienced person in the field. Including the word "best" in your search will help find such reviews, such as "best free course regression."

Read the reviews that come up, and assess the review for credibility and information content. Who is the author? What are their qualifications and experience? On what criteria are they ranking the courses? What specific criticisms or praise do they offer for each course, and are those factors relevant to your own situation?

Specializations and nanodegrees In addition to running individual courses free of charge, Coursera offers "specializations" and Udacity offers "nanodegrees" for which students must pay in order to gain the credential. These qualifications might be helpful if you have a specific job in mind.

You will support development of good content by paying for courses, but be aware that employers don't generally care as much about official certification as evidence that you can talk about the ideas from the course and that you have done some relevant work.

College credit If you want college credit for your work, then search for online offerings from a traditional university that could be transferred to your home institution. Again, you will want to check the user reviews, and also make sure you have taken the prerequisites, so that you do not waste your money.

Events and talks at your college

Campus events beyond the regular course work give you the chance to spread your wings while in a familiar environment. By attending talks in your department—such as research seminars, alumni talks, and career presentations—and going to talks outside your department by industry and non-profit leaders, distinguished academics, and government officials, you will get a sense of the big picture of our society.

These events broaden your view of what's happening in the world, and where you might you fit into it. So we recommend saying "yes," and putting the event on your schedule, the next time you hear about a special talk or panel discussion. At these campus presentations you can:

- pick up vocabulary and ideas to investigate;
- observe how people ask and respond to questions professionally;
- develop preferences on presentation style—what you do and do not like; and
- network with people from your own and other institutions, companies, and labs.

And at smaller events, it's a great idea to go up afterward and ask the speaker a follow-up question, or seek advice on career opportunities in their field. You might even gain a mentor that way.

Communication

It is difficult to make excellent presentations and to write well. Learning to communicate takes lots of practice. Below are some ways to work on your communication skills, during college and afterward.

- Courses in the humanities teach you to assess written arguments and communicate a coherent point of view, supported by evidence, in a clear and compelling way.

- Public speaking courses help you learn to consider your audience, create effective visualizations, and construct a clear, concise, and compelling story.
- Design courses teach you how to communicate your ideas visually.
- Community engagement courses and volunteer activities provide opportunities to apply your communication skills in a real setting where you can interact with local experts.
- Courses that involve teamwork and project management will force you to communicate clearly both orally and in writing with your teammates and the course supervisor.
- Courses that require a written report for a grade are also helpful in this regard.
- Presenting a poster or a talk at a conference can be a great way to practice your public speaking. You could present in a math/stat/operations research club meeting or at a regional or national conference.
- Build a professional presence on social platforms and code repositories.

Computing and coding skills

Computational tools have been vital in the quantitative sciences for a long time, and they now play a larger role than ever. Mathematical programming has become a cornerstone of data science and artificial intelligence. No matter what your job, you will need to keep up with current trends as one sector of computer science gets hot and another fades in value or usage.

We start here with a historical perspective. The take-home message is that programming languages have shifted over time and will continue to do so. It will be necessary to continue to learn and remain current.

Reprinted with permission from Randall Munroe, XKCD comic #1891.

If you were in graduate school in 1975, you probably learned to program in FORTRAN. Some of that legacy code is still in use but much of it has been converted to newer languages. Around 1985, Pascal was the language of choice. By 1995 students were learning C++ and Java. Around 2005, Python's popularity took off. By 2015, R seemed to have taken the lead for statistics, continuing a trend toward open source codes.

Operating systems and other software systems have come and gone, or adapted to survive. Algorithmic methods have trended now toward neural nets, machine learning, deep learning, statistical learning theory, and random forests, to name a few. In BIG jobs, you will likely become an expert in some of these areas but will need to learn new skills to remain current.

To give an example from the academic setting, suppose you were a high school teacher in the 1990s. You might use graphing calculators, some sort of statistical software, a graphics program, and a word processor. If you decided to go to graduate school and take a postdoctoral position in the 2000s, you would likely learn MATLAB, Maple, Mathematica, LaTeX, C++, and maybe FORTRAN and OpenMP. As a professor in the 2010s you might learn Python, R, some open source machine learning algorithms, or neural networks. You might use online project management and documentation systems.

In this example, some of the changes are the result of taking new positions, while others are driven by the development of new computational tools and software. The same will be true for you in industry or government careers.

I wish I had studied more computer science, in order to move more rapidly and efficiently from theory to implementation.

Madeline Goh, PhD, Ford

Programming overview

Basic coding is needed for any technical position. For a job based in the quantitative sciences, you do not need to be the best programmer, just able to confidently code up an algorithm to see whether it works. Of course, excellent coding skills will be a plus so don't downplay those skills if you have them. Whether you took computer science courses in college and just need to refresh your skills or have never learned to program and are learning from scratch, the reassuring point is:

If you can major in the mathematical sciences, then you can definitely learn to code.

Reprinted with permission from Randall Munroe, XKCD comic #1926.

Let's cover the high-level points before getting down to details.

1. How many programming languages? You need to be skilled in at least one programming language. Employers might specify particular languages in their job descriptions, and you might not know the ones they want. Generally, though, it is more important that you be able to transfer your coding skills and pick up a new language rather than necessarily knowing the one a particular employer wants at the moment. Data structures can be a good course to help you become a more flexible programmer.

The top-10 list of in-demand programming languages keeps changing as new languages are developed and new applications are emphasized. The good news is that programming fundamentals are transferable between languages. So you just need to be adaptable.

2. Which programming language(s)? Ideally, we would all be familiar with a high-level language good for prototyping, such as Python or R, and have some ability to do database work with SQL (pronounced "sequel"). Work using proprietary software, such as MATLAB and Mathematica, can demonstrate your prototyping ability, but for small and medium companies, open source software will be much more likely to be used. It is unlikely you will become expert at all of these, but a little knowledge can take you a long way. To get a sense of the trends in computer languages used in BIG you might consult the TIOBE Programming Community Index.

3. Why develop good coding habits? More than half the time spent programming disappears on bug finding and fixing. So it pays off to develop good coding habits. As an old saying goes, "Less haste, more speed." Start with clearly organized pseudo-code and then turn it into well-structured and well-commented programming code. You will feel slow in the beginning, but turn out fast in the long run.

4. **When is high-performance computing appropriate?** You might not have direct experience with high-performance computing, but it helps to get a general idea of how it works and when it is beneficial. Take opportunities to participate in workshops that will teach you how to port your working code to a multi-core machine. Some institutions have memberships in high-performance computing consortia so that you can apply to use computational time on a top-notch system.

5. **What are project management and repository tools?** Keeping track of who is doing what, and which tasks have been completed, is a key challenge for big team projects. So it's a plus if you have basic familiarity with version control systems for collaborative coding (such as Git or Subversion) and can post a couple of concrete examples of your projects in a web-based repository (such as GitHub)—preferably projects that show some initiative.

6. **What about spreadsheets?** Spreadsheets are pervasive in business. Many forms of data are available in tabular format, and so spreadsheets (exemplified by Excel) become the tool of choice for informing decision-makers. Excel comes with some capabilities, but when you need to do something beyond the capability of standard Excel, you can write a "macro" in Visual Basic. This practice is common in industry, where high value is placed on building a customized tool specific to a business need.

How to strengthen your coding skills

You can start today, either at your own college or studying independently online.

Courses at your college Start with a general introduction to computer science course that develops fundamental programming concepts in a language such as Java or Python. In your mathematical sciences courses, look for opportunities to do assignments in MATLAB, R, or Mathematica. Python and R are currently the two most popular programming languages in data science.

Follow up with a course on data structures and object oriented programming, and learn some C++. After that, just follow your interests....

Community organized courses Software Carpentry (which is now merged with Data Carpentry) is a non-profit organization that offers computing (and data science) basics to students and others. The courses take place on college campuses or at other institutional venues. This is a great way to learn and to network with like-minded folks. Meet-ups are also a great resource for strengthening your computing and data science skills.

Online courses   Excellent free courses are available online. Learning to code in Python and then later picking up C++ seems to work well for many people. For database work, you will need SQL.

Highly recommended courses at the time of writing include:

- Programming for Everybody (Getting Started with Python) (Coursera; by Charles Severance, University of Michigan);
- Learn to Program 1: The Fundamentals, and Learn to Program 2: Crafting Quality Code (Coursera; by Jennifer Campbell and Paul Gries, University of Toronto);
- Python for Everybody (Coursera; five course specialization by Charles Severance, University of Michigan);
- Introduction to Computer Science and Programming Using Python (EdX; by John Guttag, Eric Grimson, and Ana Bell, Massachusetts Institute of Technology);
- Introduction to Python for Data Science (EdX; by Filip Schouwenaars, Microsoft);
- Intro to Data Analysis (Udacity; by Caroline Buckey); and
- Databases and SQL (Stanford Online; three mini-courses by Jennifer Widom).

Khan Academy is another option for beginners. So is Lynda, which offers online courses on many topics. Your university might offer free access. Search for courses on C++, JavaScript, Python, and SQL.

Online offerings change all the time. The **online courses** section earlier in the chapter offers useful tips for searching.

Data science workshops

If you are interested in data science, then take a few minutes to read *How to Become a Data Scientist Before You Graduate* by Anna Schneider, available online at the Berkeley Science Review.

One pathway into data science is through a short-term, intensive workshop, sometimes called a boot camp. Established options include:

- Data Incubator (zero tuition),
- Insight Data Science Fellows Program (zero tuition),
- NYC Data Science Academy (tuition charged),
- Signal Data Science (tuition 15% of first year's salary).

Programs can be found online by searching for "data science boot camps."

If you intend to invest time and money to go through a workshop, make sure you ask a few questions before doing so. Some questions you can ask are as follows:

- How selective is your program?
- What do you do to place your graduates in their first jobs?
- How quickly do your graduates find employment?
- What is the average salary for graduates you place?
- How big is the cohort at each session?
- What is a typical training day like?
- What kind of mentoring and attention will I get?
- Give me an example of a project a trainee might work on? Do these projects come from companies associated with your program?
- Will I get training on soft skills?
- Can you give me the contact information of some of your alumni so I can talk to them?

Check the reviews Always check the online reviews of a workshop or boot camp, for example at Glassdoor.com, before applying. You want to make sure the program offers a high-quality education and well-organized professional networking. For boot camps that charge tuition, you want to be especially diligent in checking the quality of the offerings.

Business skills

If you are not yet overwhelmed with possibilities, below are business-related courses to round out your problem-solving toolbox:

- accounting,
- actuarial science,
- economics,
- finance,
- industrial psychology,
- management,
- marketing.

> *My biggest regret is not formally studying the semi-technical skills that are so critical to success in a BIG career. I learned project management, leadership and financial management on the fly. I feel I would have been far more successful if I had used seminars, books, and on-line resources to learn more systematically—which I did eventually, but much later in my career than I should have.*
>
> Allen Butler, PhD, Daniel H. Wagner Associates

Chapter 4 checklist

☐ Take some of the suggested courses, or learn independently about the topics.

☐ Plan future course work at your institution.

☐ Apply for summer programs or workshops.

☐ Consider participating on a competition team.

☐ Develop your communication skills in one of the ways suggested in this chapter.

☐ Strengthen your coding skills and create a portfolio of projects.

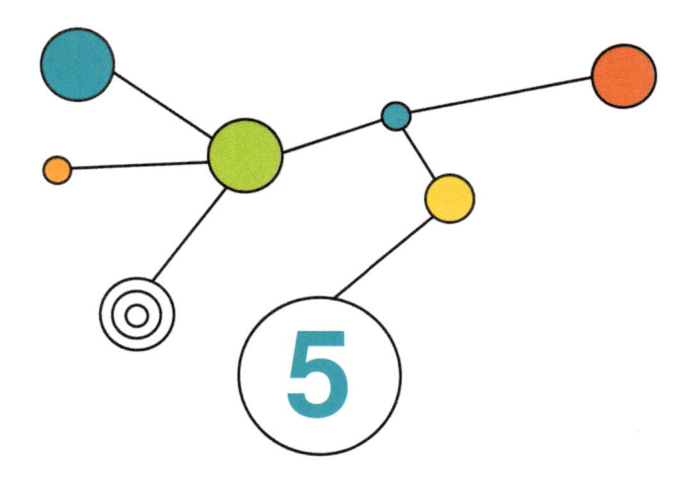

What to put in your résumé

A résumé is your marketing tool to communicate with prospective employers. It should be a concise one- to two-page document that highlights your most relevant technical and transferable skills. You should tailor the résumé each time you apply for a different position.

Sometimes you may be asked for a CV, which stands for curriculum vitae. In some cases a CV is focused and short just like a résumé, but other times it can contain your entire job, academic, and accomplishment history. So be sure you know what is being asked for, in terms of both content and length.

To get started, create a master résumé that contains all your experiences and accomplishments. Use it as the source for making a shorter, tailored résumé that highlights your most relevant qualifications, for each particular position you apply to. In most cases, your résumé will be scanned and searched automatically for keywords, so be sure to include relevant keywords from the job description.

Résumé sections

Many sections could be part of your résumé. Think about your contributions, skills you used and developed, and your significant achievements. Typical sections include education, awards, research, teaching, publications/presentations, projects, and service/outreach. You have flexibility in the choice, naming, and placement of the sections, which should all be relevant to the position you seek. While your contact information and education are usually listed first, other sections can be in any order, based on your strengths and the requirements of the position or opportunity.

Following are suggestions for each section.

Contact information   Include your name and email address and, if relevant, your telephone number and current or permanent address. You may wish to omit the postal address and phone number on a résumé, because in some cases applicants from outside the area will be screened out under the assumption they will not commute to the work site, or due to preconceived ideas about whether a person is truly interested in relocating. In those cases, think what you can do to indicate your willingness to move. For example, if your phone has a New York area code and you are searching for jobs in Illinois, you could use an address in Illinois (such as your family home, or that of a friend) to indicate you are truly interested in moving.

Summary of qualifications   This set of three to five bullet points should concisely highlight and summarize skills and experiences that relate directly to the position. Use the job description to help determine your most important qualifications.

Education   Include all institutions of higher education you have attended or are currently attending, in reverse chronological order (most recent first). Include the degree you are seeking or obtained, university name, college name, city and state of the university, and your (expected) graduation date. Including your grade point average is not mandatory. Thesis and dissertation titles, minors, course work, academic awards, and study abroad programs may also be included in this section.

Skills   Describe your tangible skills, such as coding, language, technical, and laboratory skills. State your level of proficiency, and quantify your achievements, e.g., "Proficient in C++, wrote 500 lines of code for simulating insect colony dynamics." This section is not the place to mention transferable or "soft" skills, such as communication skills.

Experience   For each experience (paid or volunteer) include your position title, the organization name and location, and dates of employment. Then create bulleted skills statements to demonstrate the skills you developed or used and the accomplishments you attained. The following formula is helpful:

- Action Verb + Details + Results (if applicable).

Begin each skill statement with a bullet point, then an action verb (see the strong action verbs listed below) to describe your relevant skill, e.g., "created," "researched," and then summarize your responsibilities, accomplishments, and projects. Describe the result of your efforts, and include quantifiable information. An example of a weak skills statement is:

Wrote code as part of a project to simulate fluid.

A stronger verb and some specific detail strengthens it considerably:

Collaborated in a team of four to efficiently model blood flow in a medical device.

Or emphasize your modeling skill and tangible accomplishments, rather than the collaboration skill, by writing:

Modeled blood flow efficiently for a medical device, generating 1000 lines of Python code in four-person collaboration.

Avoid personal pronouns such as "I," "me," "my," and make sure verbs are in the correct tense—past tense for past experiences and present tense for current experiences. List your experiences in reverse chronological order (most recent first). Consider creating specific experience sections to highlight different types of experiences, such as "Related Experience," "Research Experience," "Leadership Experience."

Research experience Include the title/type of research, and lab or department name, including the supervising faculty member. Explain the value of the work. If possible make a connection between research accomplishments and projects/topics relevant to the employer.

Focus on translatable skills you developed through research, such as scoping a project, surveying known methodologies, analyzing data, inventing new techniques, presenting results graphically, and writing reports. See the action verbs in Table 5.1 for more ideas.

Teaching experience Include titles of courses taught (not the course numbers), along with institutions and semesters taught. Do not clutter the résumé by giving every detail. It is acceptable to summarize, e.g.,

"Led discussion sections for Calculus I–III in four semesters, achieving student evaluations in top 10%."

Focus on translatable skills you developed through teaching that are applicable in industry, such as facilitating discussions, providing feedback, managing expectations, improving administrative systems, developing technical materials, and so on. Be specific about these skills, of course.

Publications and presentations   Provide a list of published works and presentations authored or coauthored (those submitted and under review), including the title, coauthors or presenters, place of publication or presentation, and dates. If included on a résumé, this list of publications should be selected based on the job description. If the papers are not relevant to the employer, then just state that you have published X papers in peer-reviewed scientific journals or conference proceedings.

Professional associations   List professional organizations in which you hold membership, including dates of your involvement and a description of your contributions, if involved beyond basic membership.

Awards and honors   List competitive scholarships, fellowships, and assistantships received, names of scholastic honors, and teaching or research awards you have received, specifically those most relevant to the position. Such honors have a half-life, so omit high school awards if you are already graduating from college, and omit undergraduate awards if you are graduating with a doctorate.

Certifications   Include earned certificates related to your field.

Grants received   Provide the names, dates, and amounts of grants you have written and received.

Choosing references

If requested as part of an application, include on your résumé the name, job title, organization name, address, phone number, and email address for three individuals. It can also be helpful to provide a brief statement describing your relationship with each reference. If included along with

Management	Communication	Research	Technical
Administered	Addressed	Analyzed	Assembled
Assigned	Brokered	Categorized	Calculated
Chaired	Collaborated	Clarified	Computed
Coordinated	Directed	Developed	Deployed
Developed	Lectured	Diagnosed	Designed
Directed	Marketed	Examined	Developed
Executed	Mediated	Identified	Devised
Organized	Promoted	Investigated	Implemented
Oversaw	Proposed	Reviewed	Modeled
Planned	Reported	Summarized	Programmed
Supervised	Wrote	Systematized	Solved

Creative	Mentoring	Accomplishment
Conceptualized	Assisted	Accelerated
Created	Coached	Achieved
Customized	Counseled	Completed
Designed	Demonstrated	Developed
Developed	Facilitated	Discovered
Integrated	Guided	Expedited
Invented	Inspired	Increased
Planned	Motivated	Invented
Proved	Supported	Launched
Streamlined	Taught	Pioneered
Structured	Trained	Transformed

Table 5.1. Strong verbs for your résumé bullet points. Follow each verb with a specific description of what you achieved.

the résumé, references should appear on a separate page that is formatted to match your résumé. However, recruiters and hiring managers know that you'll provide references on request, and they are unlikely to contact them until you make a first or second cut. You also may not want to share the contact information for your references at a career fair where your résumé might be widely circulated.

It is a great idea to prepare your reference information ahead of time, specifying the relationship with each reference person. In electronic applications it is wise to include this listing as part of the process.

Careful thought is needed about who to choose as references. If you are a graduate student, you will need to list your advisor. S/he will know you well and be able to answer questions recruiters may have.

A professor in whose class you did well might or might not be a good reference. It depends on how well they know you outside class. Did you ask questions, show curiosity, go to office hours, talk at departmental seminars or at student clubs, and so on?

If you had an internship, your supervisor there would be a good candidate for a reference. Someone with whom you did a major project is also a good candidate. The key point to keep in mind is that you want someone who can build a positive and full picture of you, so that the recruiter's decision will be an easy one.

Editing and refining your résumé

Your résumé must make a strong impression in order for you to make the cut and get to the next stage of the hiring process. So you will need to edit, edit, edit, and then edit some more! In the next pages we display a résumé first in its original form and then after editing and refinement.

Common errors

Check your résumé for these errors commonly seen in real résumés by Nicole Morgan, SPHR, SCP:

- Mailing address. Use a mailing address only if it is relevant or requested.

- Phone number. If you provide a phone number, make sure it is correct.

- Silly email addresses. Instead of using some form of name, applicants use addresses that are a disconnect for recruiters, such as roadster280@yahoo.com or lovetosurf@hotmail.com. You need an email address that is easy for the recruiter to connect to your résumé.

- Misspellings.

- An objective that is too broad or too restrictive. The challenge with an objective is that if it is too broad, it is worthless to the reader. If it is too restrictive, you could be disqualified from a position before you're really considered.

- Formatting that is difficult to read or is confusing.

Academic Job Applicant Name

myemail@somewhere.edu - postal address if relevant - webpage/url
Institution - City, State, Country

Professional Experience

Assistant Professor — Month Year – present
Department of Department Type, Institution Name

Research Instructor — Month Year – Month Year
Department of Department Type, Institution Name

Education

PhD, PhD Type, Department — Month Year – Month Year
Institution, State, Country
Advisor: Advisor Name
Dissertation: Dissertation Title

MA, MA Type, Department — Month Year – Month Year
Institution
Thesis: Thesis Title

BA, BA Type, Department — Month Year – Month Year
Institution
Honors if applicable

Honors & Grants

Honor 1 (w/ collaborators)	Granting Agency, Year
Honor 2 (relevant info)	Granting Agency, Year
Award	Granting Agency, Year
Prize	Granting Agency, Year
Scholarship	Granting Agency, Year

Research Interests

Name a few here, and limit description to a few lines.

Books/Book Chapters

Book Title (with Coauthor Names)
Book Title, Citation Info, (Year)

Chapter Title (with Coauthor Names)
Chapter Title, Citation Info, (Year)

Publications/Preprints Undergraduate co-authors indicated with *

Publication Title 1 (with Coauthor Names)
in preparation

Publication Title 2 (with Undergrad Coauthor Name*)
to appear, Journal Name, pp. start - end (Year).

Publication Title 3 (with Coauthor Names)
submitted, `arXiv:number`, pp. start - end (Year).

Conference Articles

Conference Article Title 1 (with Coauthor Names)
to appear, Journal Name, pp. start - end (Year).

Conference Article Title 2 (with Coauthor Names)
Journal Name, pp. start - end (Year).

Talks

Talk 1 (short description)	Institution, Year
Talk 2 (short description)	Institution, Year

Teaching Experience

Course 1 (short description)	Institution, Year
Course 2 (short description)	Institution, Year

Service

Activity 1 (short description)	Institution, Year
Activity 2 (short description)	Institution, Year

Notes about this résumé. This is an academic-focused résumé. The formatting is clear and concise, but the document lacks important components relevant to a BIG résumé:

- skills, including computational ones,
- project management experience (even in an academic setting),
- internships or other work in a BIG setting,
- descriptions of projects completed, with links to deliverables or code repository (or if this material is proprietary and cannot be shared, then a more general description of the type of deliverables),
- workshops and other professional development experiences,
- math modeling competitions, hackathons and other relevant team activities.

When revising your academic résumé to apply for BIG jobs, even if you do not have something to say in every category you can frame many academic experiences as comparable and relevant in a BIG context. Remember to use action verbs, which help engage the reader and convey your accomplishments.

BIG Job Applicant Name

myemail@somewhere.edu - postal address if relevant - webpage/url

Objective

You might choose to include a few sentences here about what kind of job you seek, but be careful not to be too general or restrictive. A contact in the company (or a similar company) or a career coach could be helpful as you craft this part of your résumé.

Summary of qualifications

This set of three to five bullet points should concisely highlight and summarize skills and experiences that directly relate to the position. Use the job description to help determine your most important qualifications.

Education

Under each of these degree sections, you might choose to describe projects and skills relevant to the job you are seeking. Think about what you can bring to the job from your degree program.

PhD, PhD Type, Department Completion Month, Year.
Institution, Dissertation: Dissertation Title.

MA, MA Type, Department Completion Month, Year.
Institution, Thesis: Thesis Title.

BA, BA Type, Department Completion Month, Year.
Institution, Honors if applicable.

Skills

- Computer languages: Java (basic), Python (intermediate).
- Software packages: MATLAB (incl. class projects with Signal Processing Toolbox and Simulink), Mathematica, Scikit-learn.
- Curriculum development: created learning modules for preschool through adult students.
- Languages: Spanish (fluent), French (reading proficiency).

Experience

Graduate Teaching Assistant, Department of ABC, Institution Name, Month Year – Month Year. Led recitations for three sections of Course Name. Developed and graded problem sets. Coached students as they prepared their final presentations. Explained technical ideas to audiences from general to expert, in research and educational settings, using traditional and innovative tools.

Intern, Awesome Company, City, State, Month Year – Month Year. Developed optimization algorithms for healthcare billing, as part of 15-member team. Delivered suggestions for algorithms to code developers and tested codes that were built for accuracy, speed and reliability.

Industrial Mathematics Workshop Completed intensive six-week industrial training workshop at Institute for Mathematics and its Applications at University of Minnesota, including capstone project in examining and resolving issues found in mean variance optimization.

Invited talk, Title, Conference, Year

Publications

List most relevant, with most recent first. Include GitHub or other code repository. Links are good.

Professional Associations

Association for Women in Mathematics 2011–Present. Served as Student Chapter President, 2015-17; scheduled meetings, coordinated fundraising, and maintained compliance with university-wide student group regulations. Organized an annual Sonya Kovalevsky Day with workshops for local middle-schoolers.

Society for Industrial and Applied Mathematics 2012–Present. Updated and maintained student chapter website as Student Chapter Webmaster, 2016-18. Organized industrial mathematicians to speak at student chapter meetings as Student Chapter Vice-President, 2015-16.

Awards & Honors

Honorable mention in Mathematical Modeling competition, Year

Short description of the problem and solution.

Second place in Hackathon, Sponsor, Year

Short description of the problem and solution.

Scholarship name, Granting Agency, Year

Certifications

List training experiences with links to one or two relevant examples of your work.

> **Notes about this BIG-focused résumé.** We created this résumé for early career job-seekers. The formatting is clear and concise like the academic résumé and some of the same items appear. Other items appear in a different order, and with different emphasis. A choice of narrative format or bulleted items is an individual stylistic decision.
>
> Be aware that a machine may search your résumé for keywords. If the reader is human, they may scan the first portion of each section to decide whether to put your résumé in their "yes," "maybe," or "no" pile. It is likely they will only thoroughly read the "yes" résumés. So make sure the content and formatting is easy for someone not in your field (or even in a STEM field) to read and understand.
>
> You may not have something to say in every category, but you can frame your academic experiences as comparable and relevant in a BIG context. For example, we didn't emphasize presentations or grants, but you could include them to highlight important skills, such as speaking, writing and fundraising. Remember to use action verbs, which help engage the reader and convey your accomplishments.

Reprinted with permission from Randall Munroe, XKCD comic #1911.

Your online profile

Uploading résumés, and building your network

Online job sites have profile services that provide opportunities for networking and ways to get your profile in front of recruiters. You will want to create a positive, optimistic profile page and upload a professional photo. Pay attention to which sites are being used by companies and recruiters. Include your email address in the profile, so that recruiters can easily contact you. You might consider premium paid membership for one or two of these sites while you are job hunting, to get access to all of the features.

Grow your online network by joining job site groups that focus on industry sectors or professions that interest you, and connect with alumni of your university and people whose profiles are similar to yours. Use the job sites as search engines to learn about jobs.

Project profile and code repository

Employers like to see some of your coding projects. Social platforms that use Git for version control are perfect for that purpose, so go ahead and create a profile and code repository. This repository is an extension of your résumé, much like an artist's portfolio. Remember your activity at the repository will be continuously measured and displayed. Choose wisely what you share, in order to create a favorable impression on visitors.

Curating your social media presence

Take some time to clean up your social media presence. Anytime you post information online, people can save it or pass it along. For the same reason, think carefully about what you write in emails. As the saying goes, "the e in email stands for eternal."

Employers in the U.S. can impose codes of conduct that restrict what you regard as your personal life, and they have a broad legal right to do so. You can find many news stories about people who have been rebuked, passed over for honors, and even fired because of opinions expressed on social media or in email, or because of photographs they posted of themselves or others. Please use these powerful tools carefully when you move beyond the university environment and start to share ideas and images more widely.

Your next steps

You have created a great résumé. Now it's time to apply for internships (or jobs). So turn the page and take the next steps toward your BIG career.

> **Chapter 5 checklist**
>
> ☐ Create or update your résumé.
>
> ☐ Choose three references.
>
> ☐ Curate and professionalize your social media presence.

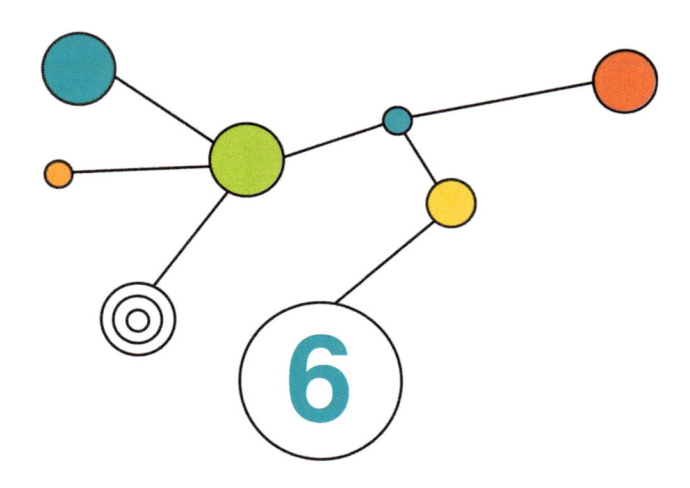

Why do an internship?

Many students tell us that an internship during their degree ended up being one of the most important preparation activities for their career.

What is an internship?

Internships come in different forms. For undergraduate and graduate students in the mathematical sciences, the most common form of internship is a paid summer job in a government lab or private company.

Other experiences count as internships too. For a mathematics undergraduate in a liberal arts college, it may be a summer job in another department during which they contribute mathematical sciences expertise to a research project. For undergraduates in engineering, the internship may come in the form of a co-op assignment where the student is placed in a company for a semester as part of meeting their degree requirements. For graduate students in statistics or operations research, an internship experience could consist of serving in a consulting group organized by the department.

What happens during an internship?

Internships provide many benefits for students and also for the companies that employ them. During an internship at a company or a lab you usually work under the supervision of a mentor. This mentor works with you to assign a project that fits well with your skills. Industry internships typically last 10–12 weeks. At the end of the internship, you usually make a presentation and deliver to the company whatever products (such as working code) you have developed. The internship is therefore an opportunity to taste what working in industry is like. Indeed, an off-campus internship often immerses you in a culture that can be very different from the more familiar academic one, and you might actually enjoy the change of pace. The perks in some workplaces are superb.

Most internship experiences involve some research and development. Thus, you will learn new skills and deploy new tools to solve your problem. Equally important is that you will sharpen your communications skills, because you will be immersed in an environment where you must communicate regularly with non-experts about technical material.

When is the best time to do an internship?

An undergraduate student in the mathematical sciences can get an internship any summer during their degree. Sometimes it is easier to get an internship than a summer Research Experience for Undergraduates (REU), because the course and skill requirements for an internship can be less specialized: companies have many ways for a student to learn and contribute to a project. You may find you have the best preparation for an internship in the summer after junior or senior year, but that will depend on your personal trajectory.

If you do find yourself with an REU opportunity, remember it can provide some of the same benefits as an internship, including teamwork, new skills, and practice communicating deliverables.

For PhD students, we recommend going for an internship during the first or second summer of graduate studies, or else toward the end of the degree, but not in the middle. (Other people recommend the middle of the degree and not the end. Individual circumstances play a big role.) For most students the first two summers are a time when you have not yet started on PhD research, and after a successful internship some students change their mind on what topic to pursue for dissertation research. An internship may in fact help define your research program.

Before you go out and seek an internship, discuss your interests with your academic advisor. It would be awkward if your advisor found out secondhand about you applying for an internship, or found out when a recruiter called them asking for a reference. Also, your advisor may be considering the research or course work needed for you to complete your degree, and so it is important to share your thoughts with them. You could share this book to help them understand the value of the internship experience for your future career goals. Together you and your advisor can determine the best time in the degree program to take an internship.

If you are a graduate student in research mode, you will not have time for thesis research or study while doing your internship. When you take a break like this from research, some catch-up time can be needed to restart the project with the same intensity. So being away from your research might have an impact on finishing your degree. On the other hand, many students experience an increase in motivation after an internship, because they have gained a career goal to work toward. Students can also gain a fresh view of their research project. Summing up, you just need to be aware of the implications for your degree progress before you apply for internships.

Benefits of internships

An internship experience prominently displayed in your résumé is helpful when you seek employment. The recruiter can learn about your internship from the host mentor (if the mentor is named as one of your references), including how well you adapted into the organization. Make sure that every experience you put on your résumé is one you can tell a brief story about—for an internship, the story should describe what you learned and what you delivered to the host.

> *From my undergraduate internships, I gained experience working at an established software engineering company. I learned how to follow best coding practices and use version control (e.g., Git, Subversion) with a large team. Beyond technical aspects, these internships taught me what sort of work culture to look out for. I also endured some of the less-desirable aspects of industry, like limited control over the project, and long delays before my code was deployed.*
>
> *My experiences in Mathematics Clinic (a year-long academic-industry capstone course) complemented these internships. As a small team, in Clinic we had much more control over the project. This meant we could spend less time with code reviews or integration, and more time prototyping or developing. Rather than adding to existing code, we had to build our repository from scratch. This process showed us some good ways (and many bad ways) to organize a project. Our weekly*

> *meetings also taught us how to best convey our findings to our liaisons. Both Clinic and my internships will provide me with valuable tools as I move into the workforce next year.*
>
> Daniel Zhang, Harvey Mudd College

Internship programs are valuable to companies too. Interns allow a company to increase their workforce (and brainpower) capacity at a relatively small cost. A good internship program is a talent magnet for the employer, because students get to see what a great company they can work for, and reciprocally the company gets to "try before they buy."

One final thing. You will be compensated handsomely by the company for your internship work. You might even score some company swag to give to family and friends.

Where to look for internships

Networking Networking through making personal connections is the best way to find an internship. If your school organizes a career fair, then go along and talk with all the companies that seem interesting. Sign up for any online job board hosted by your university because employers will recruit specifically for interns from your school. If your university has a research park or technology incubator, then check whether it offers a job board online. Talk with faculty members to see who has contacts in industry and government labs.

Your network of family and friends can also help put you in touch with internship opportunities—so go ahead and tell everyone you know about the kinds of professional opportunities you are seeking.

Smaller companies might have the ability and desire to hire an intern but not have assigned anyone in the company to actively recruit. In this case, cold-calling a contact in the company to communicate that you are available, and to discuss what kinds of skills you would bring, might generate an internship offer.

Online applications Online job sites are another way to find internship opportunities. Some of the sites are free, while others require membership in order to access the entire site. Most sites will work better for you if you build an online profile that helps train the site with information about your skill set and the type of job you seek, including any geographic preferences.

Currently, the largest job sites for North America are:

- CareerBuilder—general job board;
- Dice—tech jobs;
- Indeed—largest online job board at the moment;
- Glassdoor—anonymous reviews and salary information from employees, along with actual interview questions;
- Idealist—positions in non-profits;
- LinkedIn Premium—passive job searching;
- LinkUp—positions from company websites;
- Monster—large, general job board;
- USAJobs—federal government jobs.

Online job boards are constantly changing, and so we recommend you search for reviews of the best job boards. If you have a particular area of interest, say in data science, then you can find specialized niche job boards by searching for "best job board data science."

Students can also search directly for internships at individual company websites.

Government and professional societies The National Science Foundation in the U.S. has recently developed a graduate internship program with the national labs. In Canada, you could work with Mitacs, an organization supporting applied and industrial research in mathematical sciences and associated disciplines. In the U.K., you can seek opportunities through the Smith Institute.

Some of the professional societies run online job boards, which can be a good place to look for internships aimed at people with your disciplinary skills. Here are some active job boards:

- ASA JobWeb (American Statistical Association);
- EIMS Job Board (American Mathematical Society);
- INFORMS Career Center (Institute for Operations Research and the Management Sciences);
- Math Classifieds (Mathematical Association of America);
- MathHire (European and German Mathematical Societies);
- SIAM Job Board (Society for Industrial and Applied Mathematics).

Links are easily found online.

We also recommend Study Groups in the U.K. and Mathematical Problems in Industry workshops in the U.S., which can be great ways to gain short-term experience, similar in flavor to an internship. In these programs (usually a week in duration), students and faculty members work together to solve problems posed by practitioners in industry. These projects can turn into internships, publications, or even jobs, and can be pursued at any stage of your degree program. Read more in Chapter 8 about workshops.

Advantages of an internship

- Discover which work environments and tasks you like, and which you don't like.
- Experience how a company or lab culture differs from academic culture.
- Explore whether you want to work in a big/small or established/start-up environment.
- Compare internship experiences with other students—not all internships are created equal.
- Learn what kinds of skills you bring to the workplace.
- Work as part of a team—sometimes a large one.
- Enjoy a change of pace from your academic life.
- Learn skills that can be valuable in future jobs and on your résumé.
- Develop new skills and perspectives that you can bring back to your academic research.

A graduate student perspective

Next we hear from a graduate student about his national lab internships.

> *I did two different internships, one during the summer after my second year of graduate school and the other after my third year. (My first summer I mentored in an applied math REU program.) The first internship was at Argonne National Lab, and the second was at Sandia National Labs.*

The program at Argonne required only a CV and reference letters for the application. My CV was important in getting this internship as they sought applicants whose skills aligned well with their projects. I also had an offer with a large heavy machinery company in the same year and chose to go to Argonne instead.

I was connected to the second internship through a working group at SAMSI (a statistical and mathematical sciences institute), where I was a graduate research fellow. I was asked by another participant in the group to come and work for him.

My second internship was more productive than the first, for several reasons. I had a shorter learning curve the second summer (since national labs all have similar structures). I also started formulating the project a couple months prior to the summer and proposed my plan of work prior to arriving. I already had some familiarity with the computing infrastructure and hence didn't have to learn as much on the fly during the 12-week summer internship. By that point I also had more mathematical maturity and programming skills, both of which were needed.

As an example, during the first internship I had taken functional analysis but not used it in a research context, whereas the next summer I had used it in various projects over the course of the previous year. This enabled me to progress through the background literature more rapidly and hit the ground running.

My project at Argonne focused on sensitivity analysis in the context of simplifying the parameterization of statistical models. My project at Sandia lay at the interface of uncertainty quantification and PDE-constrained optimization. I have continued the projects at both sites, and both are progressing toward publications and potentially postdoctoral opportunities.

Joey Hart, North Carolina State University

Joey's account above describes the value of doing more than one internship. Not everyone has that luxury, but you can always compare notes with other interns to learn from their experiences and broaden your horizons that way.

The advisor's perspective

We asked Joey's advisor, who is also the director of graduate studies in the mathematics department, for a faculty perspective on internships.

Q: What timing tends to work best for internships?

A: *I recommend to do internships as soon as the student is finished with the qualifying exams; in practice, that means years 2, 3, 4. Earlier gets in the way of courses, later gets in the way of finishing on time. Of course, there are always plenty of exceptions.*

Q: As Director of Graduate Programs, do you hear concerns from faculty members about the energy students expend on internships, and that it might increase time to degree completion, or decrease the likelihood of degree completion?

A: *No. Typically, for us, internships are handled at the student-advisor level, as opposed to at the program level. So, the faculty who "allow" their advisees to seek internships are already convinced this is a good thing. We do have faculty members who regard internships with little enthusiasm; their students are unlikely to get involved in such activities. As a program director, I am trying to facilitate participation by publicizing opportunities but this is not something we can demand (unless it is made part of the degree requirements).*

Q: Do faculty members join the projects?

A: *Sometimes. Again, different faculty have different views there. Personally, I see an internship as an opportunity for the students to make new connections and learn different type of interactions and so I am perfectly happy staying in the background unless action is warranted.*

Pierre Gremaud, PhD, North Carolina State University

Chapter 6 checklist

- ☐ Determine appropriate stages for an internship during your degree program.
- ☐ Discuss internship ideas with your mentor(s).
- ☐ Plan your internship search.

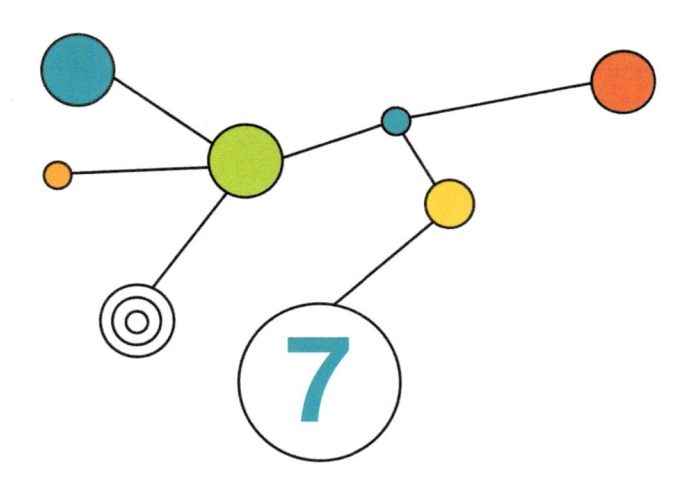

What jobs are out there?

The jobs landscape changes all the time, so in this chapter we provide some tips on how you can take a look at the job scene and decide what kinds of careers you might want to pursue. We believe that training in the mathematical sciences can provide advantages in any job, but we'll point to jobs where the skills are most directly recognizable. To navigate the jobs landscape effectively, you need to know what is out there and what jobs are best for you.

> *My mathematical training comes into play every day at work. I know that as a mathematician I've solved some of the hardest problems under the sun, and using that same analytical thinking and problem-solving process, I can approach and face and tackle any new complex problem I might find on the job.*
>
> Carla Cotwright-Williams, PhD, Social Security Administration

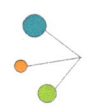

What will my job title be?

Imagine you are at a family gathering or doing outreach at a school and you ask a young person, "What do you want to be when you grow up?" It is unlikely they will respond "a mathematician!" or "a statistician!" or "an operations researcher!" Few jobs in industry come with the title of mathematician or even operations researcher, and while the title statistician occurs more often, even more common these days are titles such as data scientist or data engineer. Further, in industry it is possible (maybe even likely) that promotion will take you away from technical work as you are promoted into the management scale, so that in 10–15 years you may find yourself managing people rather than doing science.

This hidden nature of the mathematical sciences in the workplace makes internet searches for jobs (and even signing up for career fairs) somewhat challenging, and you might not know what job titles will connect you with the employers who have suitable work available for you. To acquaint yourself with the kinds of job titles you could hold, skim through the list in Figure 7.1. Browsing that list gives you ideas for career directions you might enjoy, and provides search terms you can use online.

Broad categories of careers in the mathematical sciences include research, software engineering, finance, consulting, and data science. Many mathematical sciences graduates, especially those holding advanced degrees, end up in the **research** arms of large, well-known companies. Medium and smaller sized companies that work directly with clients or contract with larger companies to provide products or services can also be good options.

A job in **software engineering** will have a targeted role of producing code that does the company's work. The algorithms driving the software often have a basis in the mathematical sciences, and employees can add value by probing why the algorithms work and how they might be improved or replaced. After developers have written code, people need to use tools (often from the mathematical sciences) to test that the code does what was intended, an acceptable percentage of the time.

The **financial** sector offers well-defined roles for quantitative analysts, who sometimes later transition into trading and managing roles. It has a reputation for being lucrative work, although often requiring long hours. Financial mathematics has strong connections to the core skills of modeling and optimization. Not everyone will feel work in finance is meaningful. The culture of the firm, and its role within the financial industry, can affect your sense of satisfaction and professional fulfillment.

Mathematical scientists can sometimes find work in **consulting firms**, where they can apply their analytical and problem-solving skills to a rapidly changing set of problems. Sometimes the work done by consulting firms is very technical. Other times the work is fairly rudimentary, but requires technical expertise not held by the client. This career works well for people who like to explore many topics and can

---

Research & Analysis
Advanced analytics specialist
Analyst
Analytics officer
Analytics scientist
Applied mathematician
Applied mathematics researcher
Consultant
Decision scientist
Mathematician
Modeler
Modeling engineer
Network optimization specialist
Operations researcher
Optimization expert
Predictive analyst
Principal scientist
Quantitative researcher
Quantitative scientist
Research analyst
Research and development analyst
Researcher
Research scientist
Scientist
Solutions architect
Staff scientist
Statistician
Technical staff
Technology officer

Writing
Associate editor
Author
Editor
Managing editor
Reporter
Staff Writer
Technical writer

Data
Analytics consultant
Analytics manager
Big data scientist
Data analyst
Data engineer
Data modeler
Data operations associate
Data processing specialist
Data scientist
Informatics scientist
Information analyst

Business & Finance
Actuary
Budget analyst
Business analyst
Business intelligence developer
Business strategy analyst
Claims specialist
E-commerce specialist
Economist
Financial analyst
Financial officer
Forecast analyst
Global pricing analyst
Investment analytics quant
Marketing analyst
Market researcher
Operations support specialist
Planner
Portfolio analyst
Product analyst
Quantitative analyst
Revenue science analyst
Risk analyst
Risk strategist
Strategist
Supply chain analyst
Trader

Engineering & Software
Engineer
Environmental analyst
Functional analyst
Geolocation engineer
Guidance and navigation engineer
Improvement engineer
Programmer
Quantitative developer
Quantitative software engineer
Reporting engineer
Research and development engineer
Simulation engineer
Software engineer
Supply chain engineer
Systems engineer

Gaming
Game designer
Game mathematician
Gamer

Security
Cryptanalyst
Cryptographer
Cryptologist
Cybersecurity specialist
Information technologist
Security analyst
Security officer

Teaching & Curriculum
Director of math tutorial curriculum
Educational software developer
Instructional designer
Math curriculum coach
Math curriculum consultant
Mathematics content specialist
Professor
Tutor

Management
Business analytics manager
Compliance manager
Manager
President
Product manager
Program manager
Project manager
Quality systems manager
Revenue manager
Supply chain manager
Team lead
Vice President

Health & Medicine
Biostatistician
Pharmacokineticist
PK/PD modeler
Quantitative pharmacologist

Figure 7.1. These job titles can help you with searches, résumés, and conversations with recruiters. Many jobs fit into more than one category, and analytical/quantitative training is a valuable asset in any job.

be flexible about dropping one line of inquiry and moving quickly on to the next project.

These days, if you are trained in the mathematical sciences, then you may find a job as a **data scientist**. Many openings exist with this or a similar job title, and almost every sector of industry is looking to strengthen their data analytics capabilities. Large, medium, and small companies, and tiny start-ups, all want to make better use of their data. Indeed, many start-up companies are founded with the explicit aim of exploiting a particular class of data, and they cannot function without data scientists.

> *I've never had the job title of mathematician, and I know very few people who do in the corporate world. It's hard to differentiate between what a mathematician does and what a software engineer does and what a computer scientist does.*
>
> Genetha Gray, PhD, People Analytics, Salesforce

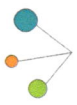

Career trajectories in industry

What does the career trajectory of a BIG employee look like? Your first job might not be a dream job, but it will certainly provide some experience and you can move on to something else you like better. The BIG job world can have mobility and opportunities to move to positions with more responsibility.

Some companies take pride in the fact that their employees stay with them for their entire working career. Companies that recruit with that goal in mind are becoming rarer, though, because the nature of an employee's work can shift over time. The employee's interests may also change and they may want to change jobs or locations for non-work related reasons. Nonetheless, good employees are valuable and companies will often go a long way to attract and retain talent.

You should regard your first job as an entryway to an industry sector. For example, if your goal is a career in the finance industry, then your first job should propel you toward your dream job in finance. Say you find a first job for a bank as a risk analyst. In that job you should take the opportunity to learn as much as possible and develop one or more areas of expertise. When you search for your next appointment, parlay your experience into a better position—better compensation, and a better fit with your interests. You can keep doing this, but you need to be seen as someone who is moving up, not laterally. You should get good recommendations from your previous colleagues, or even your employer. If you move around too much without climbing the ladder, you could be misconstrued as someone who does not fit in or is unlikely to stay anywhere.

Say you possess talents as a problem-solver and you are recognized as such by the people around you. You are so good at your work that you have developed a stellar reputation and people with challenging problems come to you. If you are content to contribute to the success of your company in this manner, then you have found your niche. However, if your dream is to manage a group or to rise to the executive suites, then you will need to seek opportunities to demonstrate leadership.

In order to have mobility, you can gently signal your aspirations. In big companies you can ask for a mentor, who can help you prepare for your next role in the company. If such a person is not available, you might consider hiring a career coach, to provide advice on how to navigate the workplace environment and get more out of your job. Either way, the point is to identify your desired career trajectory and then take steps in that direction.

Many large companies have been in the news lately regarding inequity in their workplaces: harassment, gender pay gaps, and discrimination in hiring and promotions. Find out which companies are actively addressing these problems and consider joining such an organization to become part of the force for change.

> *I left a company when I realized it did not have any promotion opportunities that offered me new ways to develop. I wanted to continue to grow technically, but all the promotions were primarily opportunities in people or project management. At that point, I knew I needed to find a company with a technical track, like 3M, where I work currently.*
>
> Catherine Micek, PhD, 3M

Start-ups versus established companies

Start-ups can be exciting places to work. On the plus side, your contribution will be felt, and you may end up playing a key role in the success of the company. At start-ups, it is frequently the case that you get involved in a broader range of areas (e.g., marketing) due to the smaller number of people at the company. And the work is typically faster paced.

You could be handsomely compensated if the company goes public or gets bought. Typically, your compensation package includes shares in the company. On the negative side, the company might fail, which would mean your shares are worthless and you must seek another job. Also, some start-ups are so small that they do not offer health benefits. Make sure you know what you are getting yourself into, and talk to others in the field.

> *Start-ups are temporary organizations used to test new business models in the wild. Most start-ups fail, but some evolve to*

Reprinted with permission from Randall Munroe, XKCD comic #1721.

become sustainable businesses solving difficult problems. The fast pace, creative freedom, and ability to break traditional business constraints attracted me to solve problems at a start-up. Customers use money as truth serum to vote on the importance of the problem and viability of the solution.

Andrew M. Webster, MS, ASA, MAAA, Validate Health

Should I create a start-up?

Do you want to create a company of your own? With the right idea for a product or service, the requisite financial backing, and the mentoring, your company might just take off. How do you start a business? While there are many resources online to tell you about the process, a very brief description follows below.

First you need to form a company. This is a relatively simple step and basically involves registering your company and determining the roles of the founders. The cost can vary depending on where you establish your "newco" and how much legal help you need. Next, depending on whether you need funding, you will need to develop a "minimally viable demo." Make sure that you have intellectual property protection. This, along with a marketing plan, will be part of your pitch to investors to show the potential of your product or service. Once funded, you will need to work hard to meet the investor-specified time lines and expectations. This fund-and-grow cycle continues until your newco exits—that is, until it gets bought, makes an initial public offering, or goes out of business.

If you are interested in entrepreneurship, you can usually find resources at your university. If you are at an institution with a business school or an entrepreneurship program, you will be able to take part in business pitches, which might even be able to connect you with venture capitalists.

> *It is rewarding being deeply involved in many early aspects of projects at a start-up. You can have a bigger impact and initiative than at a larger company.*
>
> Laura Zaremba, PhD, Groq

Consulting companies

A sector of jobs that clearly benefits from mathematical scientists is consulting. Many large management consulting companies recruit on campus. If you know former classmates who have gone on to work for such an entity, you should find out what their work is like. It may not be for everyone. For example, most of these jobs involve a lot of travel and time away from home.

Opportunities exist at information technology consulting companies particularly. Many such firms have branched into providing cloud support as well as data analytics capabilities. Their clients are companies that are not large enough to run these services in-house. These consulting companies send you out to work with their clients. Often you will be housed at the client's site. These consulting gigs could last from months to years. Sometimes the consultant is subsequently hired by the client company.

Government job opportunities and paths

The federal government in the U.S. hires mathematical sciences graduates into a broad range of government departments. Undergraduates in mathematics, statistics, and operations research can confidently apply for entry-level government positions across a wide range of fields. Higher-level internships and full-time technical positions for graduate students are available in particular at federally funded research and development organizations, most notably the national labs. A complete list of those organizations can be found by searching online.

Working conditions and retirement benefits in government tend to be good, and also stable, which cannot always be said for positions in industry.

National labs and other governmental organizations often recruit in a different way than companies. Our experience is that they prefer to get to know a candidate over a longer period, starting with a summer internship or perhaps a graduate fellowship through programs such as the U.S. Department of Energy's Computational Science Graduate Fellowship program or the National Science Foundation Mathematical Sciences Graduate Internship Program. Postdoctoral positions are also available. All these temporary positions help you decide whether working for a national lab is what you want. Note that labs do recruit directly, sometimes in person at professional society meetings. You could also make contacts through professors in your department who have national lab connections or have collaborators with connections.

Entry-level government jobs are sometimes specifically targeted to former interns from the organization. This practice is both good and bad: bad if you did not intern at the lab and want to get in, but good if it means that they might not receive many applications.

Here are some insights from a mathematical scientist in a national lab.

> *My work involves the development of numerical and computational techniques for the modeling and simulation of extremely large scale scientific problems. We do care about the theory underlying the algorithms, but the main job is to design and implement reliable algorithms that function well on our high-performance computing systems.*
>
> *Sometimes we are in competition with other government labs because we are going for the same funding, and writing proposals for the same funding calls. But we also have team projects that involve multiple institutions, some of which include other government labs. There have been team projects that have as many as ten institutions.*
>
> *Management at a lab is really different than management at a small or start-up company. At a company you may have to worry about funding on a day-to-day basis. The goal may be a product that is profitable. At a lab, in almost all cases, our projects are usually 3–5 years. This may be a bit more stable. We are not under pressure to have a product to sell. Our job is to produce algorithms and tools that scientists can use to tackle their problems. Working in a research arm of a very large stable company might be much more similar to working in a national lab.*
>
> *The career path for people with a scientific appointment is that you start as a research scientist (like an assistant professor), and*

then you can become a staff scientist (roughly an associate professor), then a senior scientist (like a full professor). We also have distinguished scientists.

We hire depending on the funding picture. We sometimes turn some of our postdocs into permanent hires. If someone has been on site as a postdoc, we know them better and know their capabilities. We have a very active internship program, particularly in the summer with undergraduate and graduate students. We even have international students and students from high school.

Esmond Ng, PhD, Lawrence Berkeley National Laboratory

Equity and your employer

Choosing a focus and location for your work means you are also choosing cultural norms that impact the working environment. These impacts can be positive, but they are sometimes negative, particularly for women and underrepresented minorities. Such norms can shift over time with changes in personnel and strategic vision in the organization, but you want to be alert from the beginning to the character of the environment you are entering, and whether your contributions and expertise are likely to be valued.

If you are reading this book as a manager, think about how you provide all new employees with a fresh start and how you can work to create equitable opportunities for professional development and promotion.

The path ahead

The career paths in this chapter all offer different affordances and challenges. The key is to find a role and an organization in which you can thrive.

Perhaps you find these possibilities exciting but you are unsure whether to make the switch from the academic life you are familiar with. Read on into the next chapter for real-life stories of students who made the leap to a BIG career.

Chapter 7 checklist

- ☐ Identify some job titles in industry and government that sound interesting.
- ☐ Learn more about those jobs.
- ☐ Decide what kind of working environment you prefer.

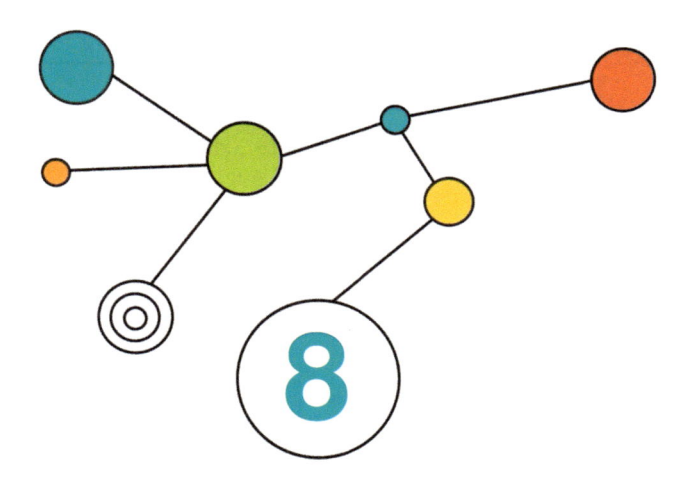

What it is like to take a BIG job

This chapter shares personal stories from the BIG Math Network blog that highlight the transition from college to the workplace. For more stories, or to contribute your own story, check out the BIG Math Network website.

From graduate school to a BIG job

by Natalie Durgin, PhD, Spiceworks

If you are nearing the end of your graduate studies and think you might like to approach the industry job market after you graduate, then I have a few tips.

Figure out how to divide your remaining time so that you are able to complete your research obligations while also devoting a percentage of time to developing a few new skills. If you sense your adviser will be supportive of this (have they had students approach industry in the past?), then let them know and get them on board. They may or may not be able to give you specific advice, but they could modify their approach during your remaining time, knowing that they don't have to get you ready to take on the academic job market. This will allow you to account for the portion of time you are spending developing new skills.

Ask the department which graduates have accepted jobs in business, industry, or government after completing their degree. LinkedIn also has a useful alumni search feature. This will give you an idea of the types of jobs to which it might be realistic to transition. At this stage, the more data you can collect, the better. Doubtless you have developed relationships and collaborations with students in your field at other math/stat departments. Ask them about students approaching industry in their departments also.

When you think about jobs in business, industry, or government, you also have the luxury of picking a location and lifestyle that suits you. Do you want to work a 60-hour week and make lots of money and live in Chicago? Do you want to work 40–45 hours a week and program and live in Portland, Oregon? Do you want to travel frequently as a consultant? There is a fit for everyone. Even if your first job doesn't fit your perfect description, get as close as you can, learn, and then keep working toward your ideal. It should be possible to achieve within a few years.

Once you know what kinds of jobs you are shooting for, work to understand the bare essentials of what you need to be an attractive candidate. Read job descriptions. Talk to your newfound industry mathematician contacts. They can help "translate" job descriptions into the few key essential skills you need to get started with the job. Job descriptions can be overwhelming and absurd, describing someone who does not exist (a "unicorn"). You don't need to go and get another degree, but you do need to be able to demonstrate that you can provide value in your chosen area.

Demonstrating value effectively will depend on which direction you have decided to pursue. In my case, I approached the company that a colleague of mine, who graduated a few years before me, had joined as a data scientist. While I still had to pass several rounds of interviews, it was invaluable that the company knew what they were getting. They knew that a pure mathematician who programmed at a certain level would be able to quickly improve because they had already seen it happen once. This meant that I needed to review the programming and algorithms courses that I took as an undergraduate, but I didn't need to be an expert on the finer points of performance engineering or details about low-level programming languages or nuances of web development.

A substitute for such a connection at your job of interest is to start a website and/or Github account that you fill with a few pet projects. Find data sets related to the ones that the company probably has, or data sets you are interested in working with. Optimize something. Build a predictive model. Build a neural net. There are countless resources that can give you ideas for pet projects. The website Kaggle has numerous data sets with problems from actual companies. Find a few projects from industry that interest you and begin building a portfolio of work that you can talk about in an interview. A company may not care how cool moduli spaces of curves are, but they will care that you built a model that improved prediction accuracy by $x\%$ above some benchmark algorithm on some benchmark data set, and that they can look at the code you wrote to do it.

I recommend choosing a job where there are at least a couple of people doing what you will be doing. This way they can quickly teach you the rest of the skills you need to be successful. For example, if you are interested in data science, choose a company where there are already a few data scientists to bring you up to speed. Choose a company that has people you enjoy working with. If you find yourself in a job where you are the first of your kind, there are always online courses and meetups. But the most direct way to absorb that information in a way that will provide fast value to your employer is to learn it from a coworker.

Good luck and have fun learning something new!

Workshops on problems in industry: the student perspective

by Hangjie Ji, PhD, University of California, Los Angeles

It is easy for graduate students to get immersed in their own narrow research area without thinking about how to use their expertise in other applications. During my PhD in mathematics at Duke University I participated in several industrial workshops where graduate students worked in teams on real problems.

These workshops provide an industry-like environment that is quite different from the usual academic setting. I attended a workshop in the Mathematical Problems in Industry (MPI) series, organized by a consortium including Rensselaer Polytechnic Institute, Worcester Polytechnic Institute, Duke University, University of Delaware, and New Jersey Institute of Technology. Engineers and scientists from industry introduced challenging problems that they face in business, and the workshop participants formed groups to tackle the problems.

The initial problems can be ill-posed in the mathematical sense, and require careful clarification and formulation to build workable models. Professors, postdocs, and graduate students work together during the week-long workshop. The team usually spends at least the first two days on resolving the confusions of the problem setup and terminology by communicating with the industry participants.

Once the mathematical model is established, various methods that we have learned are then applied to the problem, and the group members usually form smaller teams to approach the problem from different perspectives. While at first it seems a little intimidating to speak up in front of a group of senior professors, observing the way people think about and discuss problems is actually a very good learning experience, and junior mathematicians do contribute fresh insights. At the end of the week, each team presents its results, and it always amazes me how much work can be accomplished in a single week.

The other two modeling camps I participated in were designed for graduate student career development in interdisciplinary problem-solving skills. At the Graduate Student Mathematical Modeling Camp at Rensselaer Polytechnic Institute, graduate student teams get together under the guidance of invited faculty mentors to solve real problems that arise from industrial applications. These modeling camps are usually held one week prior to the MPI workshops, and part of the goal is to better prepare the grad students for the coming MPI workshop, as the week-long collaboration with team members in a friendly and productive environment quickly adjusts us to the more challenging projects we will encounter at MPI.

The Industrial Mathematical and Statistical Modeling Workshop, sponsored by the Statistical and Applied Mathematical Science Institute (SAMSI) and North Carolina State University, focuses more on data-driven industrial projects. I worked with a group of seven graduate students from different universities on a project proposed by coastal engineers from U.S. Army Corps of Engineers Coastal Lab. Our goal was to estimate the underwater bathymetry in environmental flows. Using the measurement data from several sources and statistical learning methods, we successfully incorporated the data into a wave model that leads to reasonable estimates.

I am happy with the training that I received during the workshop on technological tools like Git repository and Python, handling unfiltered messy data with simulations, and, perhaps most importantly, scientific communication skills. For instance, learning how to speak with a coastal engineer during the workshop at SAMSI turned out not to be so easy. Getting exposed to these skills, or at least being aware of the new tools people are using in industry, is very beneficial for PhD students who have interests in pursuing an industrial career after graduation. The projects during these workshops have always been nice talking points in my conversations with industrial people, and such experiences naturally contribute to a sparkling résumé for potential employers because they show interest and passion outside of one's academic research work.

I even got a publication out of the first workshop I ever attended.

Workshops on problems in industry: the faculty perspective

by Ellis Cumberbatch, PhD, Claremont Graduate University

Dr. Ji's post (above) presents a perspective from a newly minted PhD reviewing her student connections with BIG scientists and enhancing her own skills. My experience of this scene came later in my academic career as a tenured faculty member, and was twofold: as an organizer and participant in Mathematical Problems in Industry (MPI) workshops and in the Claremont Mathematics Clinic.

First, the Mathematics Clinic. These are direct connections at the Claremont Colleges for students, both graduate and undergraduate, and faculty. A project is sponsored by business, industry, or government, and is addressed by a team usually consisting of a faculty advisor and three or four students. Projects last two semesters and count for course credit. With the time available, each project team can plan a comprehensive schedule modeled on what approach might be taken in industry. Oral reports to the sponsor are made regularly and written reports are prepared for the end of each semester. The first Claremont Clinic was in 1973 and almost 300 projects have been completed since then. I have organized and participated in many projects and have joined with other faculty and our students in journal publications of results. You can find descriptions of these projects online.

Now, MPI workshops. I went to my first MPI workshop, at Rensselaer Polytechnic Institute in 1986, after five years' experience in the Clinic. This subsequently grew into the annual event that is organized by the institutions mentioned by Ji; you can search online for a list of previous conferences.

I have participated in over 20 MPI workshops, in many countries, and have co-organized four of them. Our 1986 problem turned into a short publication after some attention subsequent to the MPI week. I coedited a book consisting of 12 case studies of math problems from industry—most of them originated in an MPI workshop.

The MPI concept was started in 1968 by Oxford University applied mathematicians. (The term Study Group is used in the U.K.) It continued there successfully for a couple of decades, then expanded to Europe under the aegis of the European Consortium for Mathematics in Industry (ECMI). The ECMI website says that five to seven Study Groups are held each year in Europe, and the concept has spread to all continents. The Mathematics in Industry Information Service website lists past and future Study Groups from Europe and around the world. Some invite problems from a diversity of BIG sponsors, some restrict the applications (e.g., medicine, agriculture). This website holds a vast repository of information on the mathematics that BIG is interested in having done, as it contains past reports (or final presentations) of the work done at many MPI workshops. A student browsing through some of these can learn a lot about the modeling and math techniques that will be useful in a BIG career.

My advice to students at the graduate or undergraduate level is to adopt Dr. Ji's model of joining multiple workshops. Learning by doing can be optimal, and experiencing how other scientists think about problems enhances one's own critical and inventive facilities. My experience at MPI workshops is that the student participants benefit by seeing how an experienced modeler first tries to reduce a problem to a very simple submodel that can be understood, then the building starts. Along the way there can be many false leads that fail, but the key is to suggest alternatives again and again. The team structure brings a democratic process: a student can suggest a reason why one approach is weak, and this can then lead to its strengthening; that can be as important as the senior faculty's pronouncements. At the undergraduate level the National Science Foundation is supporting many REU summer workshops, there are internships at many national labs and other BIG companies. My advice is to apply to a number that look interesting to you—it is better than waiting tables!

Math is said to be an "ivory tower" profession, but that is far from the truth and that is why there is so much support for team projects. Join up!

Computer science skills you might not learn in school

by Lindsay Hall, Google

1. *Working in an existing codebase*

Unless you're founding your own start-up, it's highly unlikely you're going to be writing any significant code from scratch. The very first skill I had to learn was how to read and understand the existing code for my project, and how to integrate my changes into that codebase while adhering to the design patterns already in place.

Computer science courses in college tend to focus on writing code from scratch, or implementing methods in an existing class. I've never heard of a class that required students to understand and make changes to a large, preexisting codebase (although such a class might exist!). This is a crucial skill for future software developers, and one that should be stressed in college curricula.

2. *Testing code*

Writing automated tests for your code is a huge part of working in a large codebase. Tests help ensure the correctness of your code, provide information about the expected behavior of methods/classes, and protect your codebase against future regressions. Test-driven development is also a popular strategy for software development at many companies.

A few of my college courses required students to write unit tests for their code, although there was never any emphasis on testing strategies or best practices for writing unit tests. While unit testing is a big part of the test suite at many companies, other types of testing are critical as well, including integration testing, screenshot testing, and automated testing of production code using a prober or a bot. Understanding testing practices and the importance of different testing approaches is critical to working as a software engineer, especially at a large company.

3. *Writing design documents*

As mentioned earlier, I was fortunate to attend a college that placed a strong emphasis on technical communication, both written and verbal. I would say the single most important thing I do in my day is communicate with my coworkers, whether it be about code that I'm writing, or code that they're writing, or a design we're working on together.

A design document is a key component to working on a project. Before you start writing code, you need to outline your proposed changes in a format that can be easily shared with your team members and reviewed by at least one coworker. Doing these reviews before you start coding saves a lot of time and energy, since you can iterate quickly on various design ideas without having to update your code each time. Learning how/when/why to write a design doc or proposal is a skill I wish I had learned before starting work at Google.

4. *Conducting code reviews*

Many companies adhere to a code-review process where every line of code that's submitted to the codebase is reviewed by at least one other engineer. This allows for a second pair of eyes to catch bugs and suggest improvements, and also helps to spread knowledge among the team (so that at least two people know how all of the recently submitted code works). Learning to review someone else's code for correctness, style, and good design is an important skill. Also, it's important to learn how to have your code reviewed, and how to take feedback and suggestions (and when to push back on those suggestions).

5. *Working on large-group projects*

There are often many people working on a single project at any given time. In those situations, it is critical to break up the work in such a way that peoples' changes don't conflict with each other, and everyone can be productive without being blocked by someone else's changes. Learning how to parallelize the tasks in a project and coordinate across a large number of engineers is a critical skill. While some college courses encourage or require group work, most don't require students to work in groups larger than three or four people. Learning how to manage a long-term, multiperson project as part of a computer science class would be a large benefit.

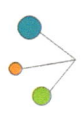

From tenure track to a BIG job

by Lucas Sabalka, PhD, Nebraska Global

I had always planned on being a professor, and that's what I became: after two postdocs and a decent rate of publication, I got a tenure track position at a research university. A career in academia has significant pluses, including the promise of tenure and thinking about interesting problems all day. However, through the course of these positions, I gradually realized two important minuses of a career in academia. One is that with academic jobs so few and far between, you typically do not get to choose where you live. My wife and I are from Nebraska and wanted to end up there, close to family and friends. The second is that research is driven by self-motivation. That's good for someone like me who is highly self-motivated, but it can also add undue stress: I was easily on track for tenure, but found myself pushing hard to make a name for myself with little recognition.

The experience that changed my career path from academia to industry was a consultant-ship. A coauthor and good friend, Dr. Josh Brown-Kramer, was working as an applied mathematician at a start-up tech company called Ocuvera, in my home town. I have an undergraduate degree in mathematics, computer science, and history, and Josh and I had competed in and won a few programming contests back in the day. I had done very little programming in the intervening years, but I had enough knowledge to pick up coding quickly. Josh put in a good word for me and got me a full-time consulting position one summer. That position allowed the company to see I was a good fit culturally and could contribute positively to their products, and it was a good opportunity for me to see what working in industry was like. A few months after the consultant-ship ended, the company extended me an offer of a permanent job. It was a difficult decision to make, but the draw of moving back home and (what was for me) the lower stress of working in industry led to my decision. I took the plunge and switched careers: from "mathematician" to "applied mathematician."

That transition was anxiety-inducing. I had prepared for many years to be in academia. It held the promise of tenure, and was familiar. Industry was scary: what if my company folded? How would I handle the different stresses?

In retrospect, I should have had more confidence in myself. The stressors are different in industry, but overall my stress levels have decreased. I have more time for hobbies, including advocacy and volunteerism. (I speak with elected officials and thought leaders about climate change and the transition to a clean energy economy.) And I now trust that I would find another job if my current job were to disappear.

Lessons learned—moving from a student perspective to an industry one

Let us summarize some key points from the above stories. Keep them in mind when starting your first BIG job!

- Take time and learn to communicate in the language of the client.
- Dig into what the problem is really about: don't solve the wrong problem.
- Be adaptive—able to quickly learn new ideas.
- Be open-minded.
- Be agile and think on your feet.
- Stay flexible and willing to switch gears into new projects, or work with someone else's idea.

Focus on what you can contribute to the organization:

- Prioritize the needs of the company and the client ahead of your own preferences and tastes.
- Recognize that BIG organizations have different compensation and reward structures than you are used to as a student.
- Internalize the 80% solution. The company does not want a 100% solution that takes two years to deliver—it wants an 80% solution in two months.

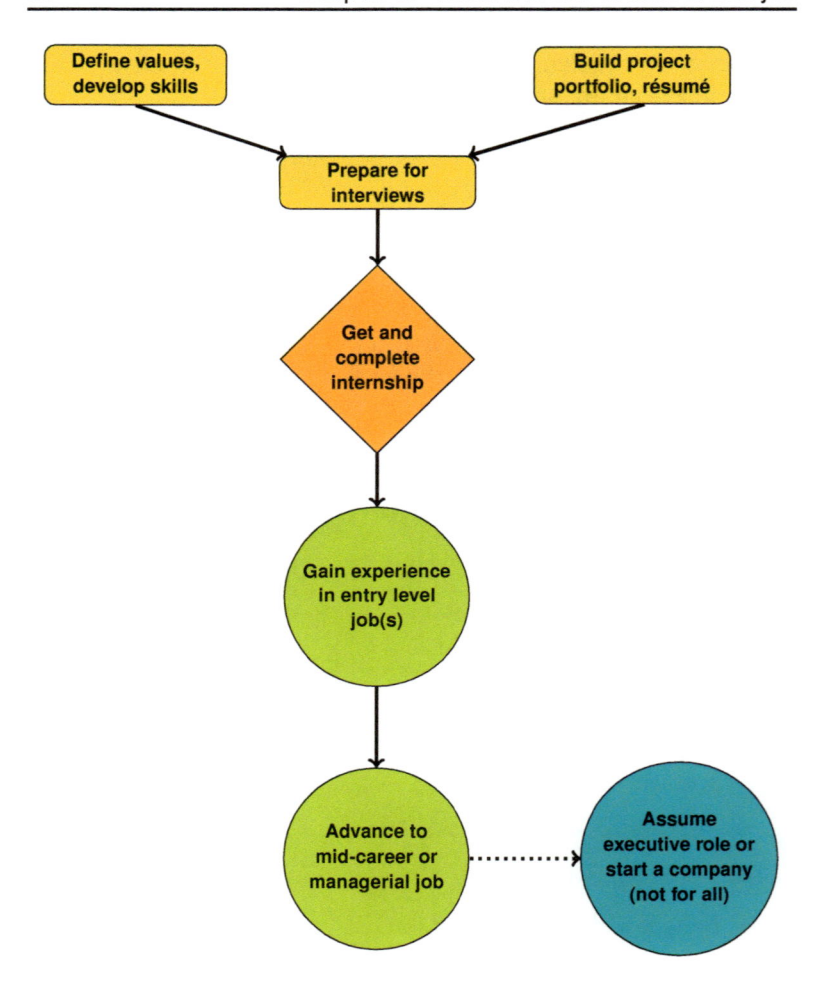

Figure 8.1. Stages of career preparation and jobs in BIG—not everyone hits every stage, or spends the same amount of time in them.

To round out the chapter, Figure 8.1 summarizes the stages of career development in BIG. Some people choose to remain in a particular stage because it suits them, while others skip to a different stage. The point is to reflect on your long-term goals and where you want to aim.

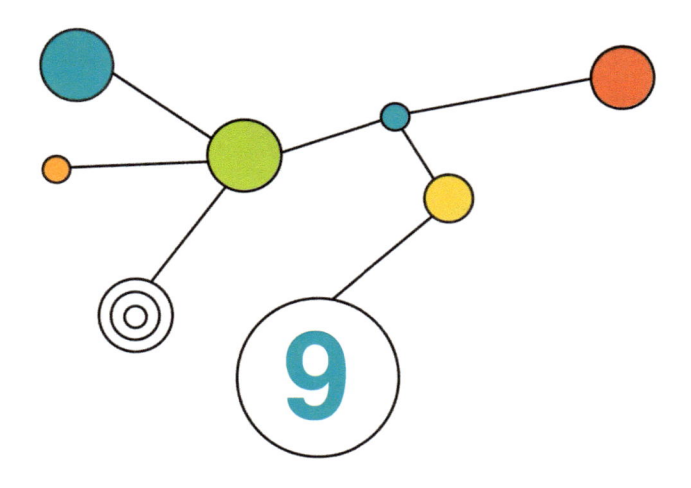

Ways to collect career mentors

Different people in your life bring different perspectives to your job training and search. You will likely find it helpful to discuss your career plans with others, even after you have managed to identify your values, refined a list of potential jobs, crafted a good résumé, and started the application process. Many people get jobs through a personal contact, so building your network of mentors can help increase your job options. Below we share ways to find and interact with career mentors.

At your university

Your academic advisor as a mentor

Have you gotten to know your advisor? At some undergraduate institutions, students may develop close relationships with advisors, but this may require some initiative on the part of the students to do more than just discuss course schedules. At some institutions, students see their advisors only rarely and don't discuss much with them. Larger schools might have professional advisors who are not faculty members, and whose knowledge about what a student can do in the workplace with a major in the mathematical sciences might be limited. (We hope this book will help those advisors, as well as students.)

Sometimes you get lucky and your faculty advisor is a source of experience and wisdom. Even if they do not have direct experience in

88   Chapter 9. Ways to collect career mentors

Reprinted with permission from Randall Munroe, XKCD comic #1917.

BIG, they can know the potential for jobs and encourage you to consider all of your options.

In graduate school, advisors have a large role to play in approving the course of study and determining when the PhD has been completed. Graduate advisors can be instrumental in helping students prepare for and attend conferences, and engage with their professional networks. They also know alumni, some of whom are working in industry.

A stigma persists (regrettably) in some PhD programs when a student lands a BIG job rather than following the academic route. PhD advisors are themselves committed to the values of academia, and you can regard it as a compliment that they would like you to join the field. Also, from the advisor's perspective, the academic reward system favors faculty members who graduate PhD students and work with them long enough to produce significant publications. Nonetheless, the majority of PhD graduates will end up in BIG careers, and advisors are increasingly recognizing this fact and encouraging students who want to go BIG.

So suppose you talk with your academic advisor about career options, and what you might like to pursue. They wholeheartedly support your ideas (you always knew they were a good egg). Even so, they might not have any relevant work experience or contacts to suggest. This book helps provide information, but all the same you need a mentor. So keep on reading....

Undergraduate or graduate director

Faculty members in these roles typically know *all* the students currently active in the program and have good knowledge of where the graduates have gone. They are excellent resources not only for advice but also for information. They usually know which professors are likely to be helpful in your quest for a BIG job. If not, then please suggest they keep some copies of this book handy in the department!

Faculty members

Ask around, talk to your classmates and professors, and find out which professors' students have ended up in industry. These professors are usually good sources of information, and can provide contacts with former students now working in industry—these students can provide on-the-ground mentoring advice.

Student clubs and organizations

While you are a student, you can usually join professional societies for a small membership fee. Sometimes it is even free. The professional societies, such as the ones that sponsor the BIG Math Network (AMS, ASA, INFORMS, MAA, and SIAM), can provide great networking opportunities and career fairs. Their publications and websites also have jobs boards, which we list in Chapter 6.

Some of these organizations sponsor student chapters, which have access to some funding for activities, and lists of speakers. You might also get ideas from other chapters about activities you could initiate on your campus. In these organizations, more senior students and alumni can serve as mentors, and groups of students can work together on skill development (through online courses, modeling competitions, and hackathons).

Career center

The campus career center typically brings recruiters and job fairs to campus, sets up off-site trips to see companies, conducts résumé reviews, provides access to self-assessment instruments that can help you identify suitable careers, and holds interview practice sessions. Undergraduates are the main audience, but the career centers are usually also open to graduate students and postdoctoral faculty.

Go to the career center early in your college career—as a freshman or sophomore. Visit once a semester, so that you get to know the staff and can incorporate them into your mentoring network.

It pays to be politely assertive with the career center. If company X is coming to interview electrical engineers, and you think you would be a good fit given your mathematical sciences or statistics degree, then you need to convince the career center that your background and experience would be valuable to the company and that they should allow you to schedule an interview. Just be realistic: if the position is not a good fit, then you don't want to waste your or the recruiter's time.

Career center staff can help you prepare for internships, conduct practice interviews, and provide guidance for the full-time job search. It often takes 6–8 months to prepare materials, figure out what you want to do, apply for jobs, interview, get hired, and start working in a full-time position. So you need to start early.

Letter writers

Most folks need a letter of reference from time to time, whether it is for a job, promotion, award, educational program, or other venture. If you are in college or graduated recently, then a faculty advisor can make a good letter writer. Later on, everyone you meet throughout your career is a potential letter writer. Your mentors can help you evaluate who would be best to ask, when it comes time to select your letter writers.

When you request a letter of recommendation, remember you are asking someone to do extra work. Most people recognize that writing references is part of their job, but it can feel to some like it falls on top of regular duties. Even if you are just asking to provide the person's name as a reference, they may get a call from someone in the human resources division, which takes time.

This means you can do some work to make it as easy as possible for your references.

- Provide a spreadsheet with the programs, contact people and information, how the reference will be submitted, and the due date. Update this periodically and let your reference know when there are changes.

- Provide your résumé or CV, and an unofficial transcript.

- Write a few paragraphs about what kinds of jobs you are seeking and why.

- Write a few paragraphs about how you know your reference. If you were in their class, explain what you got out of the class and how you distinguished yourself. You don't have to go overboard, but do give the reference some concrete things to say.

- If there is a form to fill out, fill it out as completely as possible to save your letter writer some time.

- Send friendly reminders if necessary.

- Write a nice thank-you when they have completed the submissions.

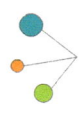

Speaking, meeting, and networking

Public speaking

If you have trouble with nervousness during interviews or while public speaking, you are not alone and help is available. You can learn to improve your speech and manage your nerves in several ways. You could join a local speaking organization, such as Toastmasters. You could take a theater, improvisation, storytelling, or public speaking course. Each of these opportunities will help you craft and present a clear, concise, compelling narrative. And the people you meet through these activities become your "confidence and presentation" mentors.

Meet-ups

Meet-ups are a great way to build your network and to connect with like-minded people. On the technical side, there are programming, data science, Python, and R meet-ups occurring at regular intervals in most metropolitan areas and on most college campuses. Meet-ups also take place for leisure activities, which you could do based on your nontechnical interests as a way to meet more people and build your network in new directions.

Alumni and friend networks

Send your résumé to a few friends who have experience with the hiring process and request their feedback. Also, ask your alma mater to connect you with alumni who can provide some career advice. For example, can you invite a graduate of your university who works at a local company to host a tour of their facility, or give a talk in your department?

Professional societies

Professional societies have job boards connecting mathematical scientists with opportunities. The boards can also be places to find internships, funding opportunities, and mentors. Join the activity groups associated with your interests and go to meet people at conferences. The larger conferences often schedule talks and panels by people in BIG jobs, and those folks will likely participate in the career fair at the conference (if offered). By making a good connection with the speaker or panelist, you gain an informal mentor.

Professional societies also offer resources for people from traditionally underrepresented groups to mentor each other, share their work, and network at conferences. They sometimes sponsor groups of talks, poster sessions, or meals at conferences. Lately, societies and organizations have also been sponsoring workshops and code-a-thons that are more BIG focused.

If you are interested in a particular company, see if any members of your professional organization work there. It is very possible that they would be willing to talk with you and help you make some connections.

Career coaches

Career coaches are paid mentors, hired as consultants to help you plan and manage your career. We have mentioned that people trained in the mathematical sciences are often pegged as "problem solvers." This can be good when you become the go-to person for people in the company who need solutions to their problems. However, you may need some help developing your own leadership skills and profile, so that your managerial potential is not overlooked.

A career coach helps you avoid such traps and determine where you want to be professionally in (say) five years' time, and then shows you how to get there. For example, suppose your goal is to lead a team working on a certain product. You must be thoughtful and deliberate in your approach, in order to send signals that you are ready for greater responsibilities, and that you want to be groomed for such a role. This is where career coaching can be of tremendous value.

Professional recruiters

It is a good idea to have mentors at every stage of your career. For mid-career people making a career change or wanting to develop targeted application materials, consulting with a recruiter can be highly beneficial. At this stage of the career, campus career offices are usually no longer available or appropriate. Friend and peer networks might be off limits until the position is obtained, because current working relationships can be negatively impacted by the news that you are seeking to move.

Dealing with a recruiter gets around those difficulties. Professional recruiters typically have regional or national contacts that can dramatically broaden the scope of your opportunities. It all helps you win at the job search, which is what the next chapter is all about.

Chapter 9 checklist

- ☐ Identify your existing mentors, and get in touch.
- ☐ Establish contact with other potential mentors.
- ☐ Join student organizations or professional societies, to expand your network.
- ☐ Visit a career center or talk with a human resources specialist/recruiter/career coach.
- ☐ Create a roster of potential letter writers. Prepare information to help them write a great letter.

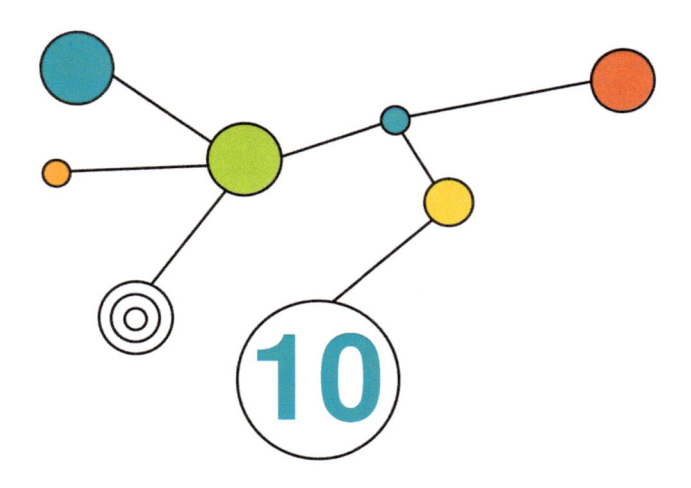

Winning at the job search

The position description when we apply for a job is never actually the work we will do on the job. So rather than applying for jobs you know you qualify for 100% according to the position description, apply for jobs based on the skill set you will bring and how you will make the job and the organization better.

Carla Cotwright-Williams, PhD, Social Security Administration

Where are the jobs?

Starting a job search can be daunting and time-consuming. You go to career fairs, use social and professional networks, and search online.

We wish we could provide magical shortcuts to find you a fulfilling job. The truth is, finding a job requires the same kind of persistence and initiative it will take to succeed after you get the job. You find a job the same way you find an internship. **Flip back to "Where to look for internships"** in Chapter 6, and develop your three-point plan of action:

1. networking,
2. professional societies,
3. online applications.

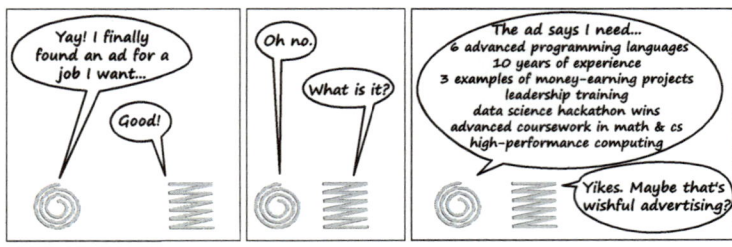

Job Ad   by Rachel Levy

Panel 1: "Yay! I finally found an ad for a job I want...." "Good!"
Panel 2: "Oh no." "What is it?"
Panel 3: "The ad says I need... 6 advanced programming languages, 10 years of experience, 3 examples of money-earning projects, leadership training, data science hackathon wins, advanced coursework in math & cs, high-performance computing" "Yikes. Maybe that's wishful advertising?"

What are you waiting for?! Go back to Chapter 6 and follow those same techniques to plan your search for a full-time job. It's a tough job, but somebody (you) has got to do it.

The online job sites listed in Chapter 6 are your allies. You should get registered, build an online profile, and spend time scoping to see what is available on each site. Your professional society's career website is worth checking too.

New opportunities are posted online daily, and so rather than trying to do your research all at once, you will benefit from regular, shorter periods of time devoted to scanning for opportunities. Keep track of which search terms gain the most traction in terms of identifying jobs you are interested in.

Getting an interview, even if it is not for your ideal job, can give you practice talking about your skills and getting over the nervousness that most people feel in interview situations. So when you get the chance, go and practice your skills in a real-life interview. Even if you are not quite right for the current opening, another position may come up later and you could get called back.

How to translate job ads into reality

Job advertisements often present a dauntingly long list of required skills and experience. **Unicorn hunting** is the name given by human resources professionals to this practice, because only one person on the planet might meet all the requirements, and it is not even likely that that person will want the job when, where, and with the benefits that are being offered. So in practical terms, no one meets the advertised requirements!

The person who gets the job is likely to be the best fit among the people who actually apply. As Nicole Morgan, SPHR, SCP, senior certified human resources professional, put it: **you don't have to be a unicorn to get a great job.**

So a job advertisement is really a wish list rather than a collection of strict requirements. When reading a job ad, categorize its requirements into three types:

(1) areas where you have experience, and reasonable expertise;

(2) skills you could easily learn if you want that type of job; and

(3) topics you might not have studied but would be willing to learn on the job.

For (2) and (3) it could be valuable to explain in the application how your current areas of expertise could help you acquire the new skill.

Do not panic if you are expert in only a couple of the requested skills. We asked some people with BIG jobs:

> If you see a job ad that lists 10 required skills, how many of them should you be expert on before applying for the position?

Much to our surprise, the answers ranged from 1 to 3 out of 10. (Respondents did say you should have familiarity with a few more of the skills if possible.) For a student used to thinking that a score less than 60% means a failing grade, it can be hard to believe that having only 30% of the required skills can be acceptable. Yet that is the case.

Women and members of other groups that are underrepresented in the field often believe they must be overqualified in order to be hired, and so might not even apply for a job unless they meet or exceed almost all the requirements. Please do not hold yourself back in this way. Our advice is to believe in yourself, put yourself out there, and see what you can achieve. Stop undercutting yourself by saying "Yes, but." Just reread this paragraph, and apply our advice to your own personal situation.

Value of diversity

When you apply for a job, remember you are a whole person, and more than the sum of your technical skills. You will bring valuable abilities and perspectives to the employer that they did not even know they needed, or thought to include in the position description.

> *The field itself is really hard regardless of gender, and as a woman, having to trailblaze on top of these difficulties is extra hard. There are some great women that have succeeded, but if you can find some trail that you don't have to blaze all by yourself, then it is going to be a little bit easier and you*

can just let your technical skills shine. I think that timing right now is good for underrepresented groups because there is more and more recognition that employers haven't been doing things correctly and need to make things more fair. We need to consider people's skills and not their gender or their race, so that's a huge plus. A lot of the way companies look is changing, and a lot of companies recognize why diversity is important for creating products.

If a person designs an algorithm and the person on the other end trying to figure out how the algorithm could be attacked thinks the same way, then you don't have any creativity. There are famous examples of companies that lost out because they didn't have diverse enough teams thinking about problems and bringing their own experiences to the table about how people actually use things and how people actually create things. In the end, most of the things you are making are for a diverse group, so you have to create in a diverse way.

Genetha Gray, PhD, People Analytics, Salesforce

Interviews

Getting an interview

If you are applying through an electronic application process, especially on an online jobs site, you are basically communicating with an algorithm. If your résumé does not contain the right search terms, it will never be read by a human. It can be hard to know what those search terms are, and if you cover your bases by peppering the résumé with large numbers of potential search terms, then you risk annoying any human who might later look at it.

The résumé will likely be filtered not only by an algorithm, but also by human resources professionals before it has any chance of being read by a manager with technical expertise. So make sure your application materials convey in clear and compelling language an accurate picture of what job you are seeking and what you will bring to the job.

For all these reasons, most job seekers do not get interviews just by uploading résumés online. You get interviews by making personal contacts.

Interview preparation

So now you have an interview. What next? To prepare for an interview you need to carry out actual research. Don't just read the company website, where everything will be described as fabulous. Read information about their competitors so you can articulate why you want to work for their organization, rather than a competitor. Use your research to develop relevant, articulate questions about the company for which you are interviewing.

Informational interviews

A key step for succeeding in the job search is the "informational interview," which is an informal conversation with someone from a company for the purpose of learning what that person does and what working for the company is like. It is not a job interview, and you do not ask for help getting hired. The exercise is useful because it informs you about the kinds of jobs out there and how people have found them. It gives you a chance to practice your elevator pitch, and tell professionals in the field about yourself in case job opportunities open up later on.

How do you find people to interview? Use your network and your mentors. Suppose you want to work as a bugle boy at Company B. Ask friends, family, neighbors, and mentors whether they know anyone working at Company B. You'll be surprised how often the answer is "yes" if you ask enough people. If not, then try using social networks to find someone at Company B who went to the same university as you, and see whether that is enough of a connection for them to agree to meet.

After you make contact and conduct the interview, you might learn that the employee does not have a background like yours, and so their pathway into the industry might not be available to you. But he or she might know someone else there whose profile is more like yours. Always conclude an informational interview by asking "who else would you suggest I talk with?"

Keys to successful interviewing

An interview is a two-way conversation in which you and a company determine whether you are a good fit for each other. It is important to prepare thoroughly. Preparation builds confidence. But realize that interviewing is a developed skill. The more you practice and prepare, the easier it will become.

Companies often conduct more than one type of interview as part of their hiring process. If you are contacted, it is fine to ask about the format of the interview. The more you know, the more prepared you will be:

- phone or video conference—this is a screening process;
- on campus—companies request to meet you for an interview, either one-on-one or group-on-one, occasionally group-on-group; usually also a screening process;
- on-site—this is where they invite you to come to their location, which provides an opportunity for more people in the company to get to know you and for you to get a better sense of the company; on-site interviews indicate serious interest.

As soon as you are contacted for an interview, do the following:

1. Learn about the company. What does it do? What are its products and services? Where does it operate? Why would you be a good fit? Read about the company's mission and values. If it has been in the news lately for reasons either good or bad, you want to know about it. For consulting firms, look for reviews from its clients.

   If possible, familiarize yourself with the technical work of the people you will meet. In order to prepare for technical interviews, you may benefit from reviewing probability, statistics, Python, and, for some jobs, SQL. Several sites mentioned in Chapter 14 have practice problems you could work on. We recommend that you work on them by yourself and then compare notes with others.

2. Articulate your values, interests, skills, accomplishments. Share your background, thesis/research/PhD work, classes, programming skills, and projects. Why should the company hire you? Know your interviewers, so that you can answer at the right level. Are they engineers, human resources staff, scientists, or managers?

3. Prepare questions about the company. What is the biggest priority for the company right now? What are areas of expansion? What is the on-boarding process if you are hired there? What are some examples of current projects? Are there technical career paths so that a person does not necessarily need to go into management to progress? What is a typical work day? What parts of the problem-solving process (requirements gathering, problem formulation/model development, algorithm development, implementation, tool transition, analysis) will the job most entail? How long do people usually stay in one part of the company?

4. Consider where you want to be placed geographically. Many companies operate at multiple locations, including internationally.
5. Be ready to talk about what you can do, and what you like to do. Talk about specific projects and abilities, rather than college courses. Always mention the value of the work you did, if possible. Prepare some technical questions to ask about what your interviewers do in their jobs.
6. Make sure that you have a narrative with each item you put on your résumé. For example, if you put down under "Experience" that you taught an undergraduate course in statistics, be prepared to tell the interviewer what was involved. Were you the sole teacher, or were you a teaching assistant for a professor? What topics were covered? How many students were in the class? What did you learn about working with other people, establishing sound procedures, and managing expectations?
7. Implicit or explicit bias might rear its ugly head during an interview. Your best approach is to be assertive and convey strongly who you are and what you are looking for in the company. Such action can prevent you from being pigeonholed from the outset. If you sense from the interview that the company will not provide a welcoming environment because of bias, and you turn down the job offer, make sure to give them feedback—let's all do our part toward achieving equity.

How did I know the job was the right fit for me? When I interviewed, it felt just like a conversation about some interesting problems and ideas.

Jeff Moulton, PhD, Google

Phone/video interviews

It is common to have a phone or video interview before you get an invitation to travel to the work site. Make sure you have arranged a quiet place with a good connection. Have a backup, whether it is phone or computer, and do not rely on the battery if possible. For video, make sure you have good lighting and pay attention to what is behind you, so that your setting is professional and not distracting to the interviewer. Make sure you have a good camera angle. Test ahead of time, ideally with a friend on the other end of a connection. Be ready 15 minutes ahead of time and download and test any videoconferencing software

Clothing choices and professionalism

The impression you make in person is crucial for winning a job. In particular, when preparing for an interview you must make decisions about clothing in which you weigh up professionalism against comfort and personal freedom of expression.

> *I feel like I should dress professionally in the way that I would want others to be dressed when they come to a meeting with me. You would never put your research paper out there or give your talk without going through and making sure that it looks correct in terms of format and trying to correct any misspellings or any bad grammar. Looking professional, acting professionally, and speaking well are the same things. You would never put up your poster with a bunch of misspellings. You take a long time to make sure everything looks right and gets your message across. You're just an extension of that.*
>
> *You can never go wrong being overdressed. At an interview they will not say anything if you are overdressed. They might say something if you are underdressed. I tend to look for shoes that are dressy but also comfortable.*
>
> *I think that minority groups always have to think about professionalism in a way the majority group often does not. So I think that seeking out those who you can talk to about it and help you make good decisions for the company culture you're in is really important, and is a little thing that can really have big results for your job.*
>
> Genetha Gray, PhD, People Analytics, Salesforce

Interview questions

Questions fall into recognizable categories.

- Behavioral—"Give me an example of…"
- Technical—"How would you solve…?"
- Intelligence/quickness—"How many marbles can you fit in a…?"

- Strengths/weaknesses—"What are your…?"
- Aspiration—"Where do you see yourself in five years…?"
- Why us—"Why do you want to work for the organization?"
- Why you—"Why should we hire you?"

Try to recognize the category of each question you are asked during an interview, and tailor your response to address the underlying issue.

Get the contact information for your interviewer so that you can send a thank-you. This might be a business card or electronic contact information file.

Let's assume you are not a unicorn, which means at least one of the interview questions will be tough. Either you don't know how to answer it, or it is on a topic with which you are not familiar. How do you respond?

First, be honest. You don't need to advertise your gaps or weaknesses, but also don't pretend you have competencies that you don't have. Second, be able to answer with what you can do and how you think about new problems you don't yet know how to solve. Talk about some things you might try and how you would go about getting up to speed. Use concrete examples of challenges you have overcome. Third, if you really liked the job and flubbed the interview, go learn the stuff and be ready the next time. Even the same employer might be eager to give another interview if you can communicate that you have picked up skills that they want.

Think about some questions to ask the interviewers that really show your interest. For example:

- How does this position contribute to the organization's success?
- How do you measure success for someone in this position?

Make sure you take note of other questions as you research the company. Here are some questions **NOT** to ask:

- Do you have positions other than this one?
- Are you interviewing other candidates?
- If I don't get an offer, can I reapply?
- Can you tell me about the health benefits and vacation plans?

These kinds of questions give the impression that you are not focused on their needs.

After the interview

Thank-you cards

After every interview, you should send each person on the hiring committee a thank-you email or handwritten note. The note should be as personalized as possible, addressed to the individual and mentioning something that you discussed with the person. There are plenty of samples online, so it is not that difficult to come up with a good one.

Follow-up contact

How long should you wait to contact them if you don't hear back right away after an interview? There is no easy answer to this. Some companies come back within hours to let you know that they are making an offer. Others take much longer, especially if they need to justify hiring you through a chain of command. The best advice is to ask at the end of the interview about when you can expect to hear from the company, and take your cue from that information. If you have a competing offer and you are interested in pursuing a position at the company, then you definitely need to contact them. At this juncture, after you have been to an on-site interview, your case is probably in the hands of the human resources department. If you got the interview through a contact in the company, you should also let that person know what is going on.

You got an offer? Now negotiate!

If this is your first job offer and it comes from a big company, the nature of the offer can be overwhelming. It will consist of many components, and you need to view the entirety as your compensation package.

Salary

The part people tend to focus on is salary. So assess whether the offered salary is competitive by searching online for "salaries at Company B" with your job title. You may need to adjust the national figures for cost of living in your location. You might not be able to trust salary information on the web.

Companies often base subsequent raises on your current salary, and so negotiating a good initial salary can yield a long-term advantage.

Ask for a signing bonus! It never hurts to ask....

Benefits

Besides salary, look carefully at the benefits package, including:

- Health insurance. This is a major expense for both the company and you. You should be offered a selection of coverage levels so that you can choose one that suits your health needs.

- Retirement plan. Does the company offer a traditional defined-benefit retirement plan? A defined-contribution plan? Or nothing at all? With any retirement plan, it is common in the U.S. for the employee to contribute a certain percentage of salary along with the company's contribution.

  Since you are a mathematical scientist who understands the power of compounding, we recommend you make the maximum possible contribution to your retirement plan while you are young.

- Life insurance and disability benefits. We advise making a careful and informed decision about these benefits. The consequences of your decision can be lifelong.

- Same-sex partner and domestic partner policies. Find out whether spousal benefits at the company extend to same-sex partners and other partnerships.

- Parental leave. What leave options are available if you become a parent?

- Publication policy. If you have a publication in preparation when you join the company, find out the company's publication policy before submitting the finished paper. It is better to ask first. Also ask about the publication policy for future work you will do for the company.

- Training and professional development. Does the company offer opportunities for further training and professional development? Most companies allocate funds for employee travel to technical conferences and workshops.

- Vacation, on-site sports facilities, on-site childcare. You may view these items as perks, but they also indicate how much the company cares about the well-being of its employees.

You might also want to inquire about company policies that do not show up in an offer, such as the on-boarding process, mentoring after you start, and so on.

Reprinted with permission from Randall Munroe, XKCD comic #1812.

Negotiating

If you are not satisfied with some aspect of the offer, then probe to see whether there is room to negotiate. If you hold a competing offer, you are in a strong position. If not, then you can still negotiate, provided you either are willing to walk away or else can develop good justifications for why you want the offer improved.

If this is an "early" offer, try to stretch the period till a decision, to see what else comes down the pike. Also see if the financial offer can be raised.

For example, suppose you are offered a job at a start-up that does not offer health insurance or retirement benefits. You could ask for more salary to make up the difference (keep a specific figure in mind). It can be more successful to make your counteroffer in writing rather than verbally, as that gives more space for the company to consider and respond. Whether in person or in writing, always articulate your needs clearly.

Women and underrepresented minority employees are often paid less than white male counterparts for equivalent positions. To counteract this phenomenon, remember the more informed you are and the more clearly you state what you want, the stronger your negotiating position. Once again, consider negotiating in writing rather than in conversation.

Hiring and retaining talent is recognized by companies as one of the most important and expensive things they must do. If they think you are a "unicorn" or even that you strongly resemble a unicorn, they will go out of their way to add you to their team.

Books and online resources about negotiating provide plenty of guidance. You can also ask friends and colleagues for advice. Our final top tips are:

- Know what you want.
- Don't be afraid to ask.
- Do your homework.
- Don't be in a hurry to finalize.
- Ask, rather than demand.

> **Chapter 10 checklist**
>
> ☐ Schedule time each week to search for and read online job ads.
>
> ☐ Prepare for interviews using suggestions in this chapter.
>
> ☐ Follow up after interviews with thank-you notes.
>
> ☐ When you get an offer, negotiate!

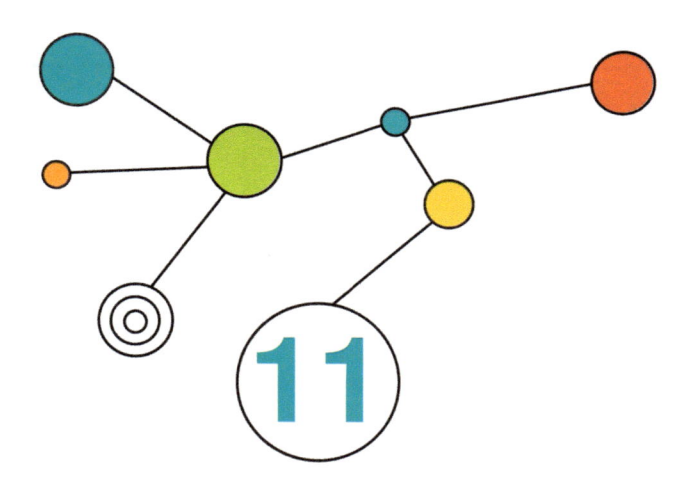

What can departments do to support BIG careers?

This chapter is designed for faculty members. Activities organized by the department can improve student readiness for industry and government careers, and in this chapter we examine a wide range of career activities along with their benefits to students.

Career activities benefit departments too, in several ways:

- students want these activities;
- student morale and future recruiting are improved;
- alumni will appreciate what you did for them, and might donate when they succeed financially in the private sector;
- the university administration will like the positive publicity from your career efforts;
- new research collaborations can be established with industry.

So we hope this chapter will help faculty members, department chairs, and directors of graduate studies to make an impact with their efforts on BIG careers.

How to use this chapter

If you can work through this material together with a couple of other people, all the better. At career workshops for faculty members you can learn about perspectives from industry, and what happens to promote careers in a variety of mathematical science departments. But even getting together for coffee with a few people from your own department can help spark local initiatives.

Current career initiatives?

Take a look at the checklist in Figure 11.1. Check off the activities your department currently does, and their frequency, in the first column on the checklist. For example, you might be conducting mock industry interviews once per semester.

Most departments are doing only a few activities from the list, and no one does all of them.

New activities—costs and benefits?

Now run through the list again and note in the second column the new activities you think might be valuable and workable in your local context. Keep an open mind at this stage—you're not committed yet!

For each activity that appealed to you, skim the relevant paragraphs below, where the idea is described in more detail along with its purpose for students or the department. Then in the second column of the checklist, classify your estimated costs of the activity (in difficulty and time) as LOW, MEDIUM, or HIGH.

Make a plan

In the third column of the checklist, select a few high-priority items to propose or plan for the near future. Jot down a target date or month, and one or two people (faculty, staff, or students) who could help you implement the activities.

Now we run through the checklist item by item.

Information and communication

Announce BIG career opportunities via email or social media

Purpose for students: receive regular alerts about opportunities both locally and nationally.

BIG Math NETWORK
Connecting Mathematical Scientists in Business, Industry, Government, and Academia

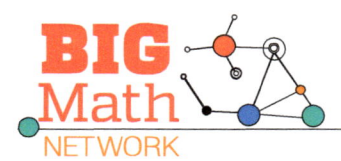

CAREER CONNECTIONS CHECKLIST

	Inventory Check if you do this	Costs (money / effort) Rate each LOW, MED, or HIGH	Action Items Check priorities
INFORMATION AND COMMUNICATION			
Announce BIG career opportunities via email listserv			
Invite recruiters to visit department			
Collect data on alumni careers and paths			
List math society job boards on department website			
Encourage students to create profiles on job sites			
Work / communicate with career services			
Connect students and faculty with BIG Math Network website			
Share math employment stats with faculty and students			
Collect (and keep current) data on local BIG employers			
Invite BIG mathematicians to advise about curriculum			
OPPORTUNITIES AND EVENTS			
Hold résumé critiquing session			
Conduct mock industry interviews			
Career panel – visitors and/or alums (could be via telecon)			
Organize on-campus internships in other departments / labs			
Sponsor coding workshops			
Sponsor trips to local BIG employers			
Advertise industrial mathematics workshops and programs			
Offer credit for industrial experience / projects (e.g. capstone)			
Invited talk by mathematical scientist in BIG career			
FACULTY			
Identify faculty with experiences in BIG			
Connect students with BIG mentors			
Incentivize faculty development (e.g. PIC Math, study groups, online courses, class auditing, boot camps, networking)			
Award prize for industrial research			
Hire BIG mathematician to teach an industry-focused course			
Provide / require public speaking courses			

CHECK OUT OUR NETWORK BLOG/WEBSITE: bigmathnetwork.org

Figure 11.1. Checklist for departmental career activities. Available for download at the BIG Math Network website (bigmathnetwork.org).

Your department needs a way to push information out to students about

- career fairs;
- recruiter visits to campus;

- industrial mathematics workshops and study groups;
- résumé and career center workshops;
- networking events;
- coding workshops;
- data science training programs;
- international student issues;
- job postings;
- society memberships, conferences, and job boards;
- research institute programs and workshops;
- internship opportunities.

An email listserv works well, at least for those students who read their email. (Many students do not.) The computer support staff in your department should be able to set up the listserv, and populate the membership list with student email addresses, updating the list every semester or year as appropriate and letting students opt out if they wish.

Email is regarded by many students as old-fashioned. Social media platforms might provide a more effective way of reaching those users, although to reach them, you will need to convince them to "follow" your feed.

Invite recruiters to visit the department

Purpose for students: connect directly with employers and discover what they are seeking.

Corporate recruiters often visit campus in conjunction with career fairs. You can hook into these networks with the help of your campus career center. Let them know what kind of employers would be a good match for your students, and they can put you in touch when recruiters come to campus.

A common format is for the recruiter to give a presentation about their company, followed by a question and answer session, and possibly individual interviews. The department's job is simply to arrange a room and computer projector and alert relevant students to the opportunity.

An alternative format is a roundtable discussion, which lets the recruiter talk more informally with a group of students. This format allow students to drop in and out as their schedules permit.

Collect data on alumni careers and paths

Purpose for students: networking and internship connections with alumni.

If you are a department chair or a director of undergraduate or graduate studies, you can benefit from keeping records of where your students go after graduation, and in the years that follow.

Students leave your program with degrees in mathematics, statistics, operations research, and data science. The ones who go straight to graduate school are usually the easiest to track. Other students take their skills into a wide variety of jobs in industry, perhaps returning later to pursue additional education. While in industry, they can be difficult to locate since companies generally do not post personnel information online.

Record keeping on alumni in industry can also be challenging because people move frequently between companies and positions. Such mobility is a standard way to advance in some sectors of the economy, in particular in the tech field. Social networks might be your best bet for keeping track of job changes for these alumni.

One additional possibility is that you might be able to coordinate with data-collection efforts in the campus alumni relations and career services offices, although only if they agree to share their data with you.

Put mathematical sciences society job boards on your department website

Purpose for students: find job listings for mathematical science students.

The society job boards are listed in Chapter 6.

Encourage students to create profiles at online job sites

Purpose for students: search for positions that match their skills, interests, and values.

The most reliable way to find a job is by networking and making personal connections. That is where students should put most of their job-hunting energy. All the same, online job boards have a role to play. See the job board listings in Chapter 6.

Most of the sites provide career advice and let users search salary information, track jobs, and upload a résumé. Students should definitely take advantage of the opportunity to create their profile at each site.

Work/communicate with career services office

Purpose for students: seek career advice from professionals.

Encourage students to take advantage of services offered at the career center. You can also invite career center staff to make presentations in the department, on topics such as résumés, networking, interviewing, and so on.

Connect students and faculty with the BIG Math Network

Purpose for students: benefit from an inspiring collection of blogs by mathematical scientists in industry and government.

Students and faculty can benefit from the BIG Math Network website, which provides resources for job seekers along with inspiring blog posts about mathematicians, statisticians, and operations researchers in industry and government.

Share employment statistics with faculty and students

Purpose for students: gain a factual understanding of career prospects for mathematical science students.

Students and faculty members are generally unaware of the true employment outlook for mathematical sciences graduates. The academic job market at the PhD level is oversupplied. Meanwhile, new opportunities are opening up in industry and government.

Chapters 2 and 14 give you resources with which to moderate an informative discussion session about the challenges and opportunities for graduates at all levels.

Collect (and keep current) data on local, regional, and national BIG employers

Purpose for students: increase the chances of getting an internship or full-time position.

We suggest assigning a careers person in your department to maintain a spreadsheet or database of employer and recruiter contacts, along with notes about preferred skill sets, interns hired, and so on. A regular part of the job should be to develop new contacts, and maintain existing ones.

National labs are a special case—considerable time and effort are required to identify good contact people inside those organizations. First, consult the National Science Foundation's list of Federally Funded Research and Development Centers and determine which ones are a good fit with the mission and expertise of your institution. Then ask faculty members in your department: do any of them have collaborators at the national lab? Do some alumni work there? Does your university have a special relationship with the lab? The vice president for research at your institution might be able to suggest lab personnel to contact.

After you have identified likely contacts inside the lab, you can inquire as to their current areas of interest, possibilities for collaboration, internships, full-time hiring, and so on. Ideally, you would arrange some visits back and forth in order to get to know them and establish an ongoing relationship.

Invite BIG mathematical scientists to provide curriculum review and advice

Purpose for department: improve the relevance of course offerings.

Undergraduate programs undergo reviews every 5–10 years. The next time your department reviews its programs, consider inviting a mathematical scientist, statistician, or operations researcher from industry or government to provide an outside perspective. Alumni are often willing to lend their experience to assist the institution.

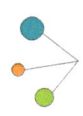

Opportunities and events

Hold résumé critiquing sessions

Purpose for students: improve application materials.

A strong résumé is essential when applying for internships and full-time positions. Students often need help refining the first draft of their résumé. Your department can steer students to the campus career center, which probably offers a résumé critiquing service, or you could invite career center staff into your department to run a workshop. You might even feel comfortable running a résumé critiquing session yourself.

For practical advice on résumés, see Chapter 5.

Conduct mock industry interviews

Purpose for students: improve interview performance.

Interviewing is a learned skill. Students get better at it with practice. So together with a couple of colleagues, or with staff from the campus career center, you could run a workshop on interviewing skills. The goal is to lead students through specific exercises to prepare them for a mock interview—then let them try out a mock interview, and give feedback on their performance.

For advice on preparing for interviews, see Chapter 10.

Organize career panels and interviews

Purpose for students: hear directly from mathematical scientists in BIG organizations.

A moderator leads an interviewee or a few panelists in a one-hour discussion of career paths in industry and government, followed by a half hour of socializing during which audience members can talk individually with panelists. During the panel, the moderator can start discussion with a few starter questions that set the scene and introduce the panelists and their backgrounds. Then, for a successful panel, the moderator must turn matters over to the audience by making clear (in a good-natured way) that he or she will wait as long as needed until audience members start asking questions.

At our own institutions, we have found it works best to be opportunistic in identifying panelists. Keep an eye out for interesting visitors to the department, and make use of your program's alumni who work in industry and government positions. To find such alumni, try talking with the undergraduate director and adviser, and the graduate director and assistant. They will know recent and older graduates with interesting career trajectories, working in government labs or industry. We encourage you to pay attention to demographics, and make good-faith efforts to include women and minority panelists wherever possible.

Panels by video link. It can be expensive to bring a visitor to campus for a career panel if they are not visiting already for other reasons. Further, a busy professional might not wish to spend 2–3 days traveling to and from your event. An economical alternative that we have tried successfully in our own institutions is a remote panel facilitated by video link. If you implement such a remote panel, we recommend that you

- make the panel 100% remote, since a mix of in-person and remote panelists will tend to marginalize the remote panelists;

- hire an audiovisual specialist to handle the logistics, including a camera and microphone trained on the audience;
- test the video and audio links in advance, and make a back-up plan.

Event checklist. Advance planning of your event is the key to success. Below are the main steps:
- Reassure the panelists that they need not prepare a presentation, and explain to them the likely composition of the audience.
- Book a room and order refreshments for after-panel socializing.
- Create an event poster and post it around the department.
- Publicize the panel on your departmental events calendar.
- Publicize the panel by mass email to a broad audience, including students, postdoctoral faculty and visitors, and tenure track and tenured faculty.
- Send individual invitations to students interested in BIG careers.
- If panelists will be on campus, then arrange individual meetings with them for students interested in their industry or company.
- Write some starter questions for the panel (see ideas below).
- Send reminders to the panelists as the date approaches.

Publicity. Whether the panel is in person or remote, you will want a good-sized audience in the room, to justify your panelists' time and effort. Publicity work is essential. We recommend sending multiple mass emails in advance of the event, to students in your department and to student clubs or groups to distribute to their members.

Mass emails tend to be ignored. Personal contact works better. To attract the audience you want, we recommend spending half an hour sending personal emails to individual students, saying that you hope to see them at the panel.

Moderating the panel. Attention to detail ensures a successful event. Start by making name cards to put in front of each panelist, or else write their name on a board behind them. Then start the panel by asking each panelist to introduce themselves and talk for a few minutes giving a brief biographical sketch: where they grew up or studied, what area they concentrated on during their undergraduate major or graduate degree, what their first job was after graduating, and the job after that.

Then ask one or perhaps two questions from the following list (not all of them), with each panelist responding in turn.

- Experiences. What experiences or events shaped your career path? What are/were the main focuses in your positions? What have you enjoyed doing professionally?
- Lessons. What is the job like for a new graduate joining your organization (team work/solo work, working hours, pathways for career development)?
- Pathways. How can a current student investigate or pursue careers in your field?

Next, encourage audience members to ask questions. Students will often be reluctant at first, but if the moderator has the patience to simply wait, someone always raises their hand and the discussion takes off.

Students might ask about longer-term career prospects in the field, work/life balance, flexibility for employees with children, and international issues such as whether the organization hires international students and sponsors them for visas or permanent residence.

To wrap things up, in the last few minutes you can ask each panelist for words of wisdom: what do they wish they had known when they were in college, and what advice do they have for current students?

Organize on-campus internships in other departments/labs

Purpose for students: gain modeling experience in an academic environment.

Students in the mathematical sciences, statistics, and operations research can contribute to scientific research groups in diverse fields such as materials science, entomology, veterinary medicine, bio-energy, public health, speech and hearing science, psychology, computer engineering, traffic engineering, and more.

To help students find placements, find out what contacts your faculty colleagues have in other departments on your campus. Do those researchers face challenges of a mathematical or statistical nature? Are they willing to host an advanced undergraduate or a graduate student in their lab for the summer or during a semester? Funding would be a bonus. Students will benefit from the research experience whether or not they get paid for their contribution.

Advertise student skills at career fairs

To advertise the skills of students in their programs, academic departments can create brochures to share at career fairs with potential internship hosts. This helps get students on the radar of recruiters, who may not realize that students from mathematics, statistics, and operations research programs can provide the technical skills their managers are seeking.

Sponsor coding workshops

Purpose for students: strengthen coding skills and develop a culture of coding among peers.

You can encourage students to participate in hackathons, software carpentry workshops, SIAM and INFORMS student chapter coding activities, online data science challenges, and so on. Your department could organize and host such events, if they are not already offered locally.

Sponsor trips to local BIG employers

Purpose for students: see an industry or government workplace in action.

Students learn a lot by getting inside the door of a local employer on an organized visit. As a department representative, you can use such a visit to convince the employer they would benefit from hiring students from your program.

Visits can be set up by cold-calling or by using the contacts of faculty and staff in your department.

Talking to technical staff at the company (rather than to human resources staff) tends to be most productive, especially if someone on the technical staff holds a PhD in the mathematical sciences, statistics, physics, or another theoretical scientific or engineering field.

Advertise industrial mathematics workshops and summer programs

Purpose for students and faculty: gain experience working in a team on a real-world industrial problem.

Industrial mathematics workshops (or study groups) typically bring together an interdisciplinary team of regular faculty members, postdoctoral faculty, and graduate students to spend a week tackling problems presented by one or more industry representatives. Summer programs and workshops provide in-depth experiences that engage students and faculty with industry problems. Examples are the IPAM RIPS program, IMA Boot Camp, and MAA PIC Math Program.

Participant perspectives on these workshops can be found in Chapter 8, written by both a graduate student and a faculty member.

Offer credit for industrial experiences/projects

Purpose for students: gain work experience while in college.

Students who undertake internships or projects with industrial partners could be offered the opportunity to gain course credit. Such co-op arrangements are common in engineering programs, for example.

Faculty initiatives

Identify faculty with experiences or interest in BIG

Purpose for department: sustain momentum on career efforts.

Team efforts are more sustainable than solo endeavors. Which colleagues could join your departmental careers team? Start by identifying colleagues who worked in industry or government, or have BIG contacts through former students or family or friends. Get the group together and pool your information, and make a plan for leveraging those BIG contacts into informational interviews, visiting speakers, career panelists, or internship hosts.

Connect students with BIG mentors

Purpose for students: get career advice from people with experience.

In the event that students do not have good access to faculty members with industry experience or other industry mentors, it can be helpful for departments to facilitate these connections. Ways include internships, participating in industrial mathematics study groups or workshops, and inviting BIG speakers into the department.

Incentivize faculty development

Purpose for department: encourage faculty members to develop expertise.

Faculty members face many demands on their time. If the department wants a colleague to (for example) learn new material in order to develop a course on the mathematics of machine learning, or supervise an undergraduate team on an industrial project through PIC Math, or create an online course, or run a computational boot camp, then some incentives might be necessary. Incentives could include a partial course release, a modest research fund, or a reduction in committee assignments.

Award prizes for industrial research

Purpose for students: reward and publicize student work on industrial problems.

When undergraduates and graduate students do internships or collaborate with a partner from industry, you want the department to benefit from their knowledge and contacts and sow seeds for future opportunities. One possible mechanism is to award an annual departmental prize for the best report, poster presentation, or talk given by a student about their internship or industrial research experience. Posters or talks could be presented at a meeting of the undergraduate student club.

Hire a BIG mathematical scientist to teach an industry-focused course

Purpose for students: understand how classroom theory relates to industrial practice.

If you know a mathematical scientist working locally in industry or government, consider inviting them to teach an industry-focused course. (It might even be possible for them to teach remotely by video link.) For example, they could teach a course on data science, or design problems in aeronautics, or challenges in internet security.

Provide/require public speaking courses

Purpose for students: improve ability to give professional presentations.

Your campus probably offers a public speaking course, which students could be encouraged or required to take in order to develop their speaking skills. They could also join Toastmasters International, which operates clubs nationwide focused on developing communication, public speaking, and leadership skills.

Suggest an invited talk by a mathematical scientist in a BIG career

Purpose for students: learn in depth about a research project in industry or government.

Mathematical scientists, statisticians, and operations researchers in industry and government often welcome the opportunity to speak about their work, both to publicize their organization for recruiting purposes and also to seek faculty research partners for their projects.

If you or your colleagues have contacts in industry, then try sounding them out about a visit. The vice president for research at your campus might be able to suggest further suitable contacts.

Does your university have connections at one of the national labs, or other federal agencies? Do any faculty members have collaborators at those institutions? Do some alumni work in one of the labs? Personal contacts will help you get a foot in the door at national labs, so it is worth investing time to build on existing connections.

Changing the culture

By implementing some of the career activities in this chapter, you will begin to change the culture in your department toward one that is more open and enthusiastic about students at all levels taking BIG internships, and going on to BIG careers.

Not all colleagues will support such culture change. For example, some might harbor concerns that their students will take longer to complete their degree if they do an internship. In response, you can advocate for students to talk with their advisor first when considering an internship, and to fit the internship appropriately into their overall academic program.

Departments with active graduate internship programs have **not** seen a lengthening of time to degree. Many students finish their studies more rapidly when motivated by the prospect of a career waiting for them in industry.

We wish you success in becoming an advocate for change.

> **Chapter 11 checklist**
>
> ☐ Download the Career Connections checklist (Figure 11.1) from the BIG Math Network and share it with your department.
>
> ☐ Prioritize a few specific initiatives.
>
> ☐ Seek support and funding from student organizations and department leaders.

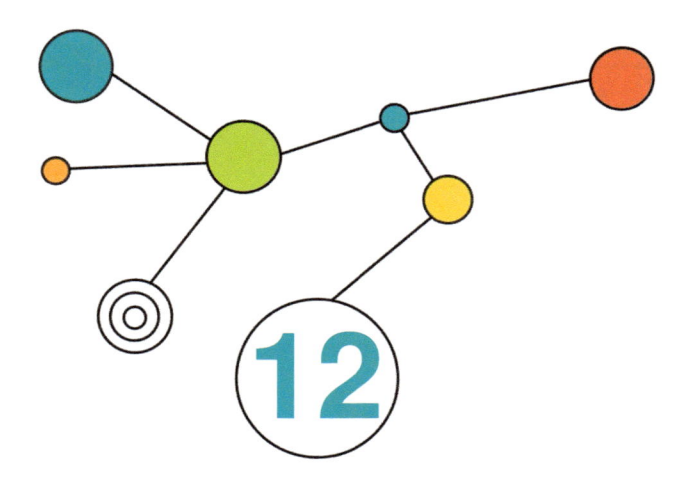

What international students and their mentors must know

This chapter addresses faculty members and department administrators who deal with international students. Most parts of the chapter are specific to students in the United States on F-1 and J-1 visas, although similar issues arise for international students studying in other countries.

Visa issues for summer internships

The following Q&A talks you through the main issues for international students who want to do an internship. Information is provided in good faith but is not guaranteed to be correct at the time of reading. Your university's international office can provide official, up-to-date guidance. In particular, students should always check in advance with the international office before accepting paid employment off campus.

Q: What visa issues must an international student address before taking a summer internship?

A: Students on F-1 visas can apply for **Curricular Practical Training (CPT)**, which allows them to accept work off campus that is integral to their major. The experience must be part of the program of study. Students on J-1 visas can get an analogous status called **Academic Training**. Contact the international office for procedures and rules.

Q: How should the department justify that the practical training experience is "integral" to the major?

A: As part of the CPT approval process, the department needs to explain how the proposed work will involve mathematical sciences content. The justification should be expressed in terms of the academic training in your program. For example, the internship might provide training in real-world applications of mathematics, complementing the theoretical work in your degree program. Here is a sample justification:

> The student's thesis research is in functional analysis, which deals with high-dimensional spaces (generally infinite dimensional) and techniques for identifying and quantifying geometric structure is such spaces. This theoretical research will be complemented by practical work during the student's internship on data science, which similarly involves high-dimensional spaces (of data) and the task of quantifying meaningful structures.

Statistics and operations research departments will generally find it easy to relate students' internship plans to research in the field.

Q: Must an international student register for a course to qualify for CPT?

A: Consult your international office about whether course registration is required for CPT status. A good option is for the student to register for an internship course. If your institution does not have such a course, then you could create one. Following is a sample catalog entry:

> MATH 497 Mathematical Internship (STAT 497 Statistical Internship). Credit: 0 hours. Full-time or part-time practice of mathematics or statistics in an off-campus government, industrial, or research laboratory environment. Summary report required. Approved for S/U grading only. May be repeated.

The requirement of a brief report at the end of the internship places accountability on the student. The report should be checked by the internship host before submission to the department, in order to avoid any intellectual property concerns.

Q: What about health insurance during a summer internship?

A: Internships generally do not include health insurance. In the summer, students can usually continue health insurance through the university by paying a fee. Students should consult the international office and health insurance office on campus for guidance.

Academic year internships

Q: Are fall or spring internships possible for international students?

A: International students may apply to work off campus during the fall or spring, subject to getting CPT or Academic Training approval. See the additional concerns below.

Concerns for academic year interns

International student interns during a fall or spring semester should find out how to

- maintain student status (full-time or part-time),
- pay tuition and fees for the semester,
- access university resources such as the library during the internship period,
- continue university health insurance.

Graduate student interns must also confirm with the department how an academic year internship will affect their university funding and tuition waiver (if they hold such funding).

Two common arrangements for international graduate students with fall or spring internships are as follows:

- Work 10 hours per week for the off-campus employer, while retaining a 10 hour per week appointment with the university in order to get a tuition waiver, fee waiver, and health insurance benefits. Register for a normal course load to maintain full-time student status with access to university resources.
- Work 40 hours per week for the off-campus employer, and pay university fees out-of-pocket, including the health insurance fee. Register for 0 hours of thesis research to maintain part-time student status and access to university resources.

Other configurations might be possible depending on university rules and the student's stage in the degree program, and on their financial resources. Students should consult with the international office to understand their options and apply for CPT or Academic Training. Students should also talk with their academic department (the graduate director or undergraduate director) to make sure they meet progress expectations during the internship period.

After graduation

Q: What rules govern work after graduation for international students?

A: Optional Practical Training (OPT) is the work authorization for F-1 students employed after graduation, and J-1 students have a similar opportunity. OPT status can begin before graduation, under some circumstances, and can be renewed for a total of up to three years for students in science, technology, engineering, and mathematics (STEM) fields. The OPT approval process can take three months or more, and so students should apply well in advance through the international office on campus.

To continue working in the U.S. after the OPT period, an H-1 visa or green card (permanent residence) will generally be needed. Before taking a full-time job, students should ask the employer whether they sponsor international workers for such visas.

International students should not assume they can remain long-term in the U.S., and are strongly encouraged to investigate career opportunities in their home countries.

U.S. students

U.S. citizens and permanent residents do not need government authorization to accept off-campus employment. Some degree programs have expectations about the number of hours of work that students may accept off campus, in order to ensure students maintain progress toward their degrees and remain eligible for financial aid.

Undocumented students are regarded as international by some institutions in the U.S., for some purposes. On the other hand, many undocumented students hold work authorization that permits internships and full-time employment. For guidance, faculty advisors can contact the university office that works with undocumented students, especially regarding work authorization.

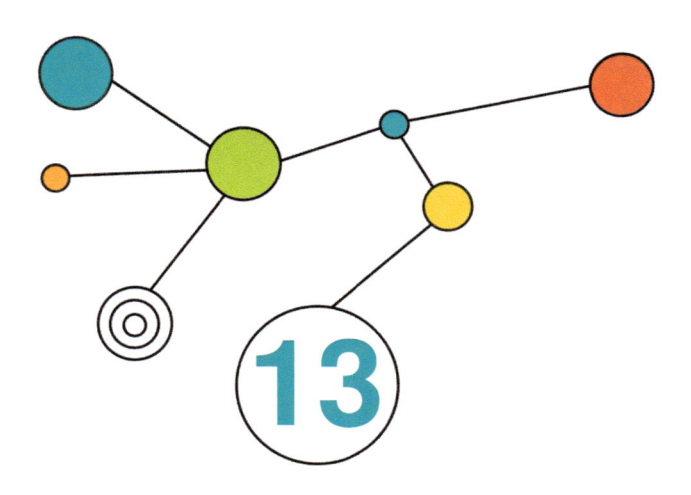

Ways to establish and mentor internships

This chapter is intended for academic institutions and companies that have not yet established an internship program. We start with the story of a well-established industrial mentoring program.

A mentor's perspective on internships

by Lalitha Venkataramanan, PhD, Schlumberger–Doll Research

I have mentored about ten summer interns in 20 years at Schlumberger. They have mostly been 1st–4th year PhD students, from various universities in the U.S. and some from abroad. The internships were well paid and lasted between 3 and 6 months. Since most of the interns worked in the field of image and/or signal processing, they all had a good background in physics and/or electrical engineering, with a good, solid mathematical background. Most of the internships have led to a patent memo, and in one case to a commercial algorithm.

Before the interns arrive for the summer, I do a lot of preparatory work and formulate a clear set of objectives I would like the intern to achieve. Sometimes, I start working on the intern's problem before s/he arrives. This gives me a rough idea of how the problem might work out and allows me to anticipate other issues we might face in the project.

Setting the expectations correctly in the beginning is very important. Most internships start with my giving the interns a few papers on the subject and encouraging them to read and come back with questions within a week. Their questions help me understand what they do and do not know and how best to help them (and therefore help me) during the internship.

A critical parameter that varies widely from one student to another is the level of mentor involvement. Most students are self-driven. However, some need more day-to-day assistance than others. I make it a point to formally meet each student at a specific time every day, even if it is for a short duration. I get a sense of what they are working on and encourage them to share their work, even if the contribution is small. This gives me an opportunity to help them navigate some local problems such as "I don't know how to access this file" or "How do I approach this problem?"

Most interns get excited when they realize that their ideas could potentially lead to an invention disclosure and patent application. This means they work very hard. It is helpful for interns to socialize with other interns and have a social circle to hang out with after work and during weekends. I encourage most of my interns not to work on weekends (unless they want to).

All of my interns now work in industry. After their PhD or master's, they often ask for recommendations, which I am happy to provide. I have learned something from each of my interns—a new algorithm, a new way to code something, some aspect of mathematics that I was unfamiliar with, or a math problem I did not know how to solve. I hope that they have learned something from me as well and have returned to their schools to finish their PhDs with a sense of a project well done!

Host questionnaire

The lessons in the previous section from Lalitha Venkataramanan's experience are clear: whether you are a faculty member seeking to place students with a company or a company representative reaching out to faculty members, preparation and supervision are the keys to a successful mentoring experience.

Companies recruiting students from a particular university department can help the academic institution identify suitable candidates by completing the following host questionnaire.

1. Name of organization and host unit.
2. Name and position of mentor(s) at the company.
3. Phone and email contact information.
4. Brief paragraph describing likely project topic(s).
5. Course work and skills expected of the employee.
6. Other skills desired or expected of the employee (e.g., programming skills, subject matter knowledge).
7. Desired internship or job period.
8. Restrictions (e.g., security clearance, intellectual property issues, safety training).
9. Other information for the student (e.g., whether project is open ended, any existing contacts in the mathematics, statistics, or operations research departments).
10. References and subject matter entry points for the student (e.g., survey articles, tutorials, Wikipedia entries).

Academic departments can develop their own versions of this questionnaire for local host companies to use. The result will be an informative compilation of the types of positions available to students.

Preplanning for mentors

A clearly defined and communicated mentoring plan can help make internships productive for everyone involved, rather than becoming managerial time sinks.

Some students will come to the host with prior experience on team projects that required project management, documentation, and report writing. Most will further develop such skills during the internship. So if you are supervising an intern from mathematics, statistics, or operations research for the first time, you may benefit from considering the following questions.

- To what extent do you plan to define the scope of work before hiring the intern and seeing their skill set and interests?
- What will you tell the student early on to make sure they do not use proprietary ideas (such as unpublished thesis research with an advisor) or proprietary software (that your company does not want to license)?

- How can you help the student understand the culture of the company (level of formality, communication methods, working hours, and so on)?
- How independent do you want the intern to be? Do you want to be accessible to them throughout the day? Once per day? Weekly?
- Could it be beneficial to hire two or more interns so that they can work together and keep each other engaged and on track?
- How much will you need to check the intern's work for it to be valuable to the company? How can you structure your expectations so that the intern can do some of the proof of concept, quality control, and testing work?
- What kind of documentation will you need in order to make use of the results after the internship is over? Can that documentation be produced regularly in an ongoing fashion, in case the internship is cut short (due to illness, for example)?
- How regular will the working hours be? How should the intern report their use of time?
- What kind of checkpoints can you establish so that the intern receives feedback before settling too firmly on one solution path?
- What opportunities for small and large presentations can you create, so that the interns practice communicating their ideas? Presentations also prompt interns to create graphics that help you later when reporting on the success of the internship to managers within the company.
- How can you provide mechanisms for the intern to get up to speed on the problem, and on relevant techniques and technology (tutorials, reading, videos, literature review)?
- How can you avoid scope creep (or mission creep), which could result in a mess of unfinished and not-useful results?

Project scoping and management

Project scoping

Project scoping is an essential aspect of an internship. It could help a host to talk with people who have previously mentored interns, to see what range of accomplishments they have observed in the past. Some mentors try to identify three possible projects ahead of time, so that the eventual project can be matched to the intern's skills and interests.

The ideal internship project has no real downside for the host and considerable upside potential. In particular, the project should not lie on a critical path for the company: if the project fails, the company should not be harmed. At the same time, ideally you can identify a project where the possible reward is high: where, if the solution works, the company could really benefit.

As one of our colleagues likes to say, mathematical sciences students are best at the "R" side of R&D, where they can focus on research and prototyping rather than development and implementation. At the end of an internship, these students will provide pseudocode or prototyping code for solving the problem, rather than fully optimized production code.

Departments can help the hosts with project scoping by identifying and communicating settings in which mathematics, statistics, or operations research have significant and interesting impact, especially at the graduate level. Students will generally want to solve a problem that could be of genuine importance to the company, and they will want the work to involve creative aspects. For example, a company might hold a trove of customer data that is not being exploited, and could scope out an internship project that applies data science techniques to seek actionable insights from the data.

Teamwork

We generally recommend that interns work in a team. This structure provides opportunities for the interns to bounce ideas off each other and maintain each other's spirits and momentum. Of course, merely placing people in a group does not ensure good outcomes. Groups need to establish expectations for behavior, create regular opportunities to check in and make adjustments to the work plan, and follow systems for dealing with unproductive or counterproductive team members. The internship host should help the team establish these expectations and systems for working together.

Academic settings typically reward individual work, and so interns might not be accustomed to cooperating and compromising, depending on others, and being dependable themselves. Sometimes team members will not get along. Sometimes one member will be too strong-willed and certain that their way is the right way. Ideally, team members will contribute complementary strengths and communication styles. The host can help team members identify their strengths and roles within the team.

Often it works well to pair up a PhD student who focuses on theory with an undergraduate or master's student who focuses on coding and implementation. The master's student might be happy to continue full time with the company after the internship, as could undergraduates or PhD students at the end of their programs.

Project management

In BIG organizations, projects usually involve large groups of contributors. Mentors will get the most benefit from interns by setting clear expectations and plans for checking in and reporting. The intern needs to understand the larger goals of the project, what their piece of the project is and who it serves, and the milestones and deliverables.

Further, the working relationship between mentor and intern will go best if there is a match in expectations about how often, for how long, and in what way they will touch base. Options include in person, by teleconference, email, or project management system.

Students can learn a lot from paying attention to project management, such as the pathway from prototyping to implementation, and how to estimate the time needed to complete tasks along the way. They can also learn the importance of logistical issues, such as run time for code. Through paying attention to project management, students who have been taught in their academic work to seek optimal solutions will learn instead to produce practical and feasible solutions.

Intellectual property

Company and university intellectual property

If the intern is a graduate student, the internship or job mentor may benefit from asking the student to talk with the faculty advisor and establish clear boundaries around intellectual property. A student in an internship might be inclined to use techniques developed with their advisor that are not yet published and probably should not be shared with the company without the advisor's permission. Or the student might share what they learned during the internship with their advisor or other academic collaborators. To avoid such misunderstandings and inadvertent release of proprietary information, the company should clearly delineate what information the student is permitted to take back to their work outside the company.

In addition, it is helpful if the company can explicitly state what information the student may share when they go on the job market or are giving a talk about their internship experience. For example, the student and mentor could craft a paragraph for the student's online profile or could prepare slides for a short presentation. The student can prepare a one-page report with visuals that can be preapproved for external consumption.

If the intern is a paid employee of a company, the intellectual property issue is clear because the company owns everything. Neither the intern nor the home institution can lay claim to an invention created while the intern is employed by a company. The intern will be asked to sign an agreement to this effect before the internship begins.

The situation can be more complicated if the internship takes the form of a research assistantship and is supported through a sponsored project with a company. Usually in this case, the university will negotiate the intellectual property agreement. The student needs to be aware of the disclosure procedures, and inform both the school and the company what the agreements are with each. Failure to do so can jeopardize the student's standing with the company, university, and project.

Student intellectual property

If you are a student who develops a concept while an employee of the university, as either a teaching assistant or research assistant, and even if the idea is developed on your own time, you should proceed cautiously. The university might still claim partial intellectual property. You can clarify your position by talking with an official at the university who is responsible for patents; start by contacting the university office of "technology commercialization" or similar title. They might suggest filing a patent, in which case you will be the author but the university could be the owner. If the university makes money on the patent through licensing, you may get a share of the income. And you certainly get bragging rights—"I have a patent!"

If after describing your discovery the university is not interested in pursuing protection, or if the circumstances are such that the university cannot lay claim to ownership of the intellectual property, then you are free to pursue protection on your own. It is costly to hire a lawyer to file a patent. A good approach in the U.S. might be to file a provisional patent (not examined by the patent office and therefore easy to get), which protects you for one year. During this protection period, you might want to establish a company and develop a so-called minimally viable product (a demo) to secure venture funding. Once that financial backing is in place, you could file a patent.

Reporting

Proper reporting will guarantee the results of the internship are summarized usefully for the host and also the student and their institution.

Technical report for host

The host should give specific guidance on the desired format of the report, since many students are unfamiliar with writing technical reports. Reports usually describe the goals and milestones of the project, the minimal viable product, and estimates of what percentage of the project components have been completed and what resources would be needed to improve the solution by a specified amount.

Report and presentation for academic institution

A report to the institution can help the company recruit future interns and employees, and help the academic institution talk about the contributions made by their students during their internships.

Ideally, the student will return from the internship with the company's written approval to share material about the internship experience that can go on their online profile or the department website: where they worked, how they found the position, what the working environment was like (solo, or team), a broad overview of the problem, the general techniques used to solve the problem, and the results of the project (such as the recommendation to senior managers, or prototype code). Also useful would be slides for a 10-minute public presentation to the department, possibly adapted from internal presentations conducted during the internship but now vetted for external release.

14

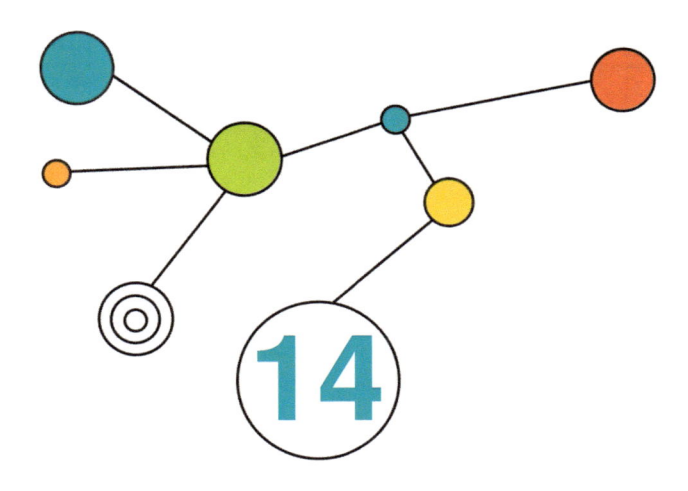

Where to learn more

We hope you have enjoyed this book and been inspired to start a career in BIG or help others to plan their own BIG career. For more information and inspiration, we recommend the following resources.

Books

Case in Point 9: Complete Case Interview Preparation by Marc Cosetino (Burgee Press, 2016). This book covers interview skills, particularly for consulting jobs, and career case studies.

Putting Your Science to Work: Practical Career Strategies for Scientists by Peter S. Fiske (American Physical Society, College Park, MD, 2012). This free online booklet includes interview advice and a summary of stereotypes held by academics about business people and by business people about academics.

Great Jobs for Math Majors by Stephen E. Lambert and Ruth J. DeCotis, 2nd edition (McGraw-Hill Education, New York, 2005). Aimed at undergraduate mathematics majors, this book offers practical job-hunting advice and anecdotes from professionals in a variety of fields.

She Does Math! Real-Life Problems from Women on the Job edited by Marla Parker, Classroom Resource Materials Series (Mathematical Association of America, Washington, DC, 1995). Career histories are presented for 38 women, along with examples of mathematical problems solved by them in the course of their jobs.

How to Become a Data Scientist Before You Graduate by Anna Schneider, *Berkeley Science Review* (online). The author provides a lively, short, practical guide to the ten things she did during graduate school in biophysics to prepare for a career in data science.

101 Careers in Mathematics edited by Andrew Sterrett, 3rd edition (Mathematical Association of America, Washington, DC, 2014). Over 101 brief profiles are included of people who majored in the mathematical sciences and now work (mostly) in government and industry. Updated regularly; next update scheduled for 2019.

Strong Interest Inventory (available online). This career self-assessment tool is one of several on the market that you could consider using as you evaluate your personal strengths and career interests. Your campus career center might offer such assessment tools free of charge.

Mathematical Problems in Industry workshops

Some long-running activities in this area include

- Graduate Student Mathematical Modeling Camp (U.S.),
- Mathematical Problems in Industry workshops (U.S.),
- Mathematics in Industry Study Groups (Europe).

Information about these programs and how to get involved can be found online. The Mathematics in Industry Information Service website is a particularly useful resource for activities in Europe and internationally.

Online courses

Check sites like Coursera, EdX, DataCamp, and Udacity for the latest courses on subjects such as data science, the technical interview, quantitative analyst courses for finance, and machine learning. See the tips in Chapter 4 about how to find high-quality courses.

Online resources

ASA Bachelor's Survey — includes lists of job titles and specific companies that hire statisticians with bachelor's degrees
AAAS careers — website
BIG Math Network — blog posts about BIG careers
Career Cornerstone Center — career center focused on STEM careers
From PhD to Life — career transition stories
Jobs on Toast — website

PhDs.org — career website
SIAM articles on mathematical sciences careers — at the SIAM website
Versatile PhD — website

Coding practice

HackerRank.com — coding challenges
Project Euler — mathematically inspired coding challenges
Cracking the Coding Interview: 189 Programming Questions and Solutions by Gayle Laakmann McDowel, 6th edition (CareerCup, 2016)

Search terms

Career advice is abundant online, and of abundantly variable quality. To browse the latest advice, try the following search terms:

- What not to do in an interview.
- Interview tips and advice.
- Building a great résumé.
- Public speaking.
- Negotiating and job contracts.
- Teamwork.
- Visualization of quantitative information.

Use your judgment about the qualifications of the person giving advice. Is their advice sensible, and corroborated by other sources?

Words of wisdom

> *My advice to those on the job market is: don't give up. The job market is difficult these days. Academic jobs are on the decrease. Government and industry jobs are available, but can take time to get because of the extensive security clearance process for many positions in science, technology, engineering, and mathematics, and also just because of the overall supply and demand for everyone. So be open. Be patient and prepare yourself for your dream job. It may take a little time.*
>
> Carla Cotwright-Williams, PhD, Social Security Administration

Index

accounting, 43
actuarial science, 43
alumni, 113, 115
AMS, American Mathematical
	Society, vii, xi, 61, 89
artificial intelligence, 32
ASA, American Statistical
	Association, vii, xi,
	34, 61, 89, 136

bias, 101
BIG, Business, Industry,
	Government, vii, ix,
	1, 2, 136
BIG Math Network, xi, xii

capstone, xii, 30, 54, 59, 111
career center, 8, 18, 20, 89–90,
	112, 114–116, 136
career coaches, 92
career fair, 17, 20, 60, 91, 95,
	111, 112, 118
clothing, 102
code repository, 37, 40, 55
coding, 34, 37, 38, 55, 68, 78,
	80, 82–84, 100, 112,
	119, 129, 131, 137
coding interview, 137
coding theory, 31
combinatorics, 27, 32
communication skills, 36, 49,
	58, 80, 83
community college, 12

competitions, 34
computer science, 31, 32, 37,
	40, 82
consulting
	management, 68, 73, 135
contests, 34, 84
	modeling, 34
cryptography, 31
Curricular Practical Training,
	123
CV, curriculum vitae, *see also*
	résumé 45–56, 63, 90

data analytics, 69, 73
data science, 32, 34, 41, 69, 78,
	112, 136
design
	experimental, 32
	visual, 37
differential equations, 27, 33
discrete mathematics, 27
discrimination, 71, 101
diversity, 97–98

economics, 43
electromagnetics, 32
elevator pitch, 21, 22, 99
entrepreneurship, 72
equity issues, 71, 75, 97–98,
	102, 106, 116

finance, 43, 68, 136
financial mathematics, 68

139

gender pay gap, 71
government careers, 73, 136
graph theory, 27, 32

harassment, 71

image processing, 31, 33, 127
INFORMS, Institute for Operations Research and the Management Sciences, vii, xi, 25, 34, 61, 89, 119
intellectual property, 72, 124, 129, 132–133
international office, 123–126
international students, 112, 118, 123–126
internship, 57–64, 91, 112, 114, 118, 120, 122, 127–134
internship course, 124
interview, 79, 89, 96, 135–137
    coding, 137
    informational, 20, 99
    mock, 116
    on-site, 21, 100
    phone, 21, 100, 101
    preparation for, 99
    questions at, 102–103

job advertisements, 78, 96–97
job board, 60–61, 113
job descriptions, 78, 95–97
job market
    academic, 9–15

keywords, 19, 45

learning
    deep, 38
    machine, 32, 38, 120, 136
    statistical, 38

linear algebra, 28
linear programming, 27

M3 Challenge, MathWorks Math Modeling Challenge, vii, xi, 34
MAA, Mathematical Association of America, vii, xi, 61, 89
machine learning, 32, 38, 120, 136
management, 43
marketing, 43
Mathematics Clinic, xii, 59, 81
mechanics
    continuum, 31
    fluid, 31
mentors, 4, 12, 19, 22, 34, 36, 42, 49, 57, 59, 62, 71, 72, 80, 87–92, 99, 105, 120, 123, 127–134
modeling
    mathematical, 28, 34, 79–82
MPI, Mathematical Problems in Industry, 62, 79, 81, 119, 136

national lab, 4, 62, 73, 115, 122
negotiating salary and benefits, 104–107, 137
networking, 55, 95, 112
numerical analysis, 28

online courses, 35, 136
Optional Practical Training, 126
OR, Operations Research, vii, 25, 27, 32, 34

Index

PIC Math, Preparation for Industrial Careers in Mathematical Sciences, vii, 120
position
    adjunct, 15
    postdoctoral, ix, xii, 3, 4, 9–10, 12, 14, 63, 74, 89, 117
    tenure track, 3, 4, 9–15, 84–85, 117
postdoctoral faculty, ix, xii, 3, 4, 9–10, 12, 14, 63, 74, 89, 117
probability, 29
programming languages, 37, 39, 78
psychology, 43
public speaking, 37, 91, 121, 137
publications, 48

quantitative analyst, 68, 136

recruiters, 92
references, 48–50, 90
résumé, 45–56, 90, 98, 112, 115, 137
résumé keywords, 19, 45
REU, Research Experiences for Undergraduates, vii, 34, 58
reviews, 35, 42

salary and benefits, 4, 8, 19, 61, 113
    negotiating, 104–107
scientific computing, 28
self-assessment tool, 18, 89, 136
senior thesis, 30
SIAM, Society for Industrial and Applied Mathematics, vii, xi, 2, 61, 89, 119, 137
signal processing, 24, 33, 127
simulation, 33
social media, 55, 110
software engineering, 68, 83
spreadsheets, 40
start-up, 1, 21, 62, 69, 71–74, 82, 84, 106
statistics, 29
stochastic processes, 30
Study Group, *see also* MPI, Mathematical Problems in Industry 62, 81, 119, 136
summer programs, 34

teaching experience, 47
time series, 33

unicorn hunting, 96

venture capital, 73
visa issues, 123
volunteer activities, 37